NON-WELL-FOUNDED SETS

CSLI
Lecture Notes
Number 14

NON-WELL-FOUNDED SETS

Peter Aczel

Foreword by Jon Barwise

CENTER FOR THE STUDY
OF LANGUAGE
AND INFORMATION

CSLI was founded early in 1983 by researchers from Stanford University, SRI International, and Xerox PARC to further research and development of integrated theories of language, information, and computation. CSLI headquarters and the publication offices are located at the Stanford site.

CSLI/SRI International
333 Ravenswood Avenue
Menlo Park, CA 94025

CSLI/Stanford
Ventura Hall
Stanford, CA 94305

CSLI/Xerox PARC
3333 Coyote Hill Road
Palo Alto, CA 94304

Library of Congress Cataloging-in-Publication Data

Aczel, Peter, 1941–
Non-well-founded sets.

(CSLI lecture notes ; no. 14)
Bibliography: p.
Includes index.
1. Axiomatic set theory. I. Title. II. Series.
QA248.A28   1988 511.3'22   87–17857
ISBN 0–937073–21–0
ISBN 0–937073–22–9 (pbk.)

To my mother and father,
Suzy and George Aczel

Let E be a set, E' one of its elements, E'' any element of E', and so on. I call a *descent* the sequence of steps from E to E', E' to E'', etc. .... I say that a set is *ordinary* when it only gives rise to finite descents; I say that it is *extraordinary* when among its descents there are some which are infinite.

–Mirimanoff (1917)
Les antinomies de Russell et de Burali-Forti
et le problème fondamental de la theorie des ensembles

Contents

Foreword

To my way of thinking, mathemtical logic is a branch of applied mathematics. It applies mathematics to model and study various sorts of symbolic systems: axioms, proofs, programs, computers, or people talking and reasoning together. This is the only view of mathematical logic which does justice to the logician's intuition that logic really is a field, not just the union of several unrelated fields.

One expects that logic, as a branch of applied mathematics, will not only use existing tools from mathematics, but also that it will lead to the creation of new mathematical tools, tools that arise out of the need to model some real world phenomena not adequately modeled by previously known mathematical structures. Turing's analysis of the notion of algorithm by means of Turing machines is an obvious example. In this way, by reaching out and studying new pheonemona, applied mathematics in general, and mathematical logic in particular, enriches mathematics, not only with new theorems, but also with new mathematical structures, structures for the mathematician to study and for others to apply in new domains.

The theory of circular and otherwise extra-ordinary sets presented in this book is an excellent example of this synergistic process. Aczel's work was motivated by work of Robin Milner in computer science modeling concurrent processes. The fact that these processes are inherently circular makes them awkward to model in traditonal set theory, since most straightforward ideas run afoul of the axiom of foundation. As a result, Milner's own treatment was highly syntactic. Aczel's original aim was to find a version of set theory where these circular phenomena could be modeled in a straightforward way, using standard techniques from set theory. This forced him to develop an alternative conception of set, the conception that lies at the heart of this book. Aczel returns to his starting point in the final chapter of this book.

Before learning of Aczel's work, I had run up against similar difficulties in my work in situation theory and situation semantics. It seemed that in order to understand common knowledge (a crucial feature of communication), circular propositions, various aspects of perceptual knowledge and self-awareness, we had to admit that there are situations that are not wellfounded under the "constituent of" relation. This meant that the most natural route to modeling situations was blocked by the axiom of foundation. As a result, we either had to give up the tools of set theory which are so well loved in mathematical logic, or we had to enrich the conception of set, finding one that admits of circular sets, at least. I wrestled with this dilemma for well over a year before I argued for the latter move in (Barwise 1986). It was at just this point that Aczel visited CSLI and gave the seminar which formed the basis of this book. Since then, I have found several applications of Aczel's set theory, far removed from the problems in computer science that originally motivated Aczel. Others have gone on to do interesting work of a strictly mathematical nature exploring this expanded universe of sets.

I feel quite certain that there is still a lot to be done with this universe of sets, on both fronts, that there are mathematical problems to be solved, and further applications to be found. However, there is a serious linguistic obstacle to this work, arising out of the dominance of the cumulative conception of set. Just as there used to be complaints about referring to complex numbers as numbers, so there are objections to referring to non-well-founded sets as sets. While there is clear historical justification for this usage, the objection persists and distracts from the interest and importance of the subject. However, I am convinced that readers who approach this book unencumbered by this linguistic problem will find themselves amply rewarded for their effort. The AFA theory of non-well-founded sets is a beautiful one, full of potential for mathematics and its applications to symbolic systems. I am delighted to have played a small role, as Director of CSLI during Aczel's stay, in helping to bring this book into existence.

JON BARWISE

Preface

This work started out as lecture notes for a graduate course called "Sets and Processes" given in the Mathematics Department at Stanford University in the period January—March 1985. In fact some of the material was handed out week by week as the course progressed and was then tidied up to form the first draft of the book. Jon Barwise suggested that this should appear in the CSLI lecture notes series, and I eagerly and gratefully accepted his suggestion. I think that the idea in both our minds at the time was that a few more weeks of work on the notes would put them in suitable shape for publication. This idea turned out to be mistaken and several years have passed. I hope that the book is somewhat better than it would have been had I finished it according to plan.

In many respects the book follows the structure of the original draft. In particular there are exercises scattered through the text which were put in for a variety of reasons. Some of them are essential for what follows, but only involve routine arguments. Others make points that are less essential. Parts one and two contain a revised version of the original draft. Part three contains work related to material that was given in the original lectures, but in much more detail. In addition I have added an appendix, Notes Towards a History, in which I attempt to present the historical information I have gathered in working on the book. A surprising range of people have written about non-well-founded sets, often in ignorance of each other. I have tracked down some of this work. No doubt there are yet more discoveries to be made here.

In an attempt to give the book a wider readership I have included an appendix, Background Set Theory, in which the reader can find out what are the set theoretical notation and ideas that are needed to read the book. The reader is recommended to start out directly with chapter one which requires only a limited familiarity with set theory. Later chapters require more familiarity and the reader should consult the appendix as necessary.

This book would never have appeared without the initial suggestion and continual encouragement of Jon Barwise. He has displayed great patience and also imagination in finding new reasons why the book ought to be finished. More importantly he has discovered an area of application for non-well-founded sets, Situation Theory, which promises to make these sets of wider interest than I could have envisaged when I first became interested in them. That application is barely treated in this book. For that the reader should refer to some of his publications. In particular the book, *The Liar,* written by John Etchemendy and Jon Barwise, can usefully be read in conjunction with this book. That book gives an elegant intuitive treatment of the anti-foundation axiom and makes an interesting application of it to the philosophical problem of the liar paradox.

As usual there are a variety of people and institutions who had a hand, sometimes unforseen, in the completion of this book. The excellent physical and intellectual environment of Stanford University and CSLI was an important stimulus for me. I am grateful to the System Development Foundation who funded my visit to CSLI during the period October 84–April 85. It was there that I first encountered the marvelous TEX typesetting system. I learnt to use LATEX, the package of TEX macros designed by Leslie Lamport, by using it to write the original drafts of this book, and have continued to use it since with great enthusiasm. Donald Knuth deserves the gratitude of many people like myself for his creation of the TEX system. After visiting Stanford I took up a 3 month SERC research position at Edinburgh University during the summer of 1985. I am grateful to Gordon Plotkin for organising this, and also to him and Robin Milner for discussions with them on concurrent processes while I was there. I came to learn that the notion of a concurrent process was a good deal more complex and subtle than I had thought when I first started to think about the notion and its relationship to non-well-founded sets. Robin Milner's work on SCCS was the direct cause for my original interest in non-well-founded sets.

Some of the writing of this book took place while I was on leave from the Mathematics Department at Manchester University, with a research position at the Computer Science Department of Manchester University, during parts of 1986 and 1987. I am grateful to ICL for the funding of this arrangment.

I would like to thank Emma Pease at CSLI, for her work in translating the LaTeX files of this book into the appropriate TeX files used in the CSLI Lecture Notes Series, and to Judy Boyd at Manchester University, for her help with some of the LaTeX typing. Dikran Karagueuzian managed the production of this book and I am grateful for his advice and patient assistance at the various stages of the book's progress.

I thank my wife Helen for her continual encouragement. She shared with me the ups and downs involved in the completion of this book. Finally there is lovely Rosalind. She cannot really be blamed for the further final delay that marked her first six months of life.

Manchester
24 December 1987

Introduction

A non-well-founded set is an extraordinary set in the sense of Mirimanoff.* Such a set has an infinite descending membership sequence; i.e. an infinite sequence of sets, consisting of an element of the set, an element of that element, an element of that element of that element and so on ad infinitum. What is extraordinary about such a set is that it would seem that it could never get formed; for in order to form the set we would first have to form its elements, and to form those elements we would have to have previously formed their elements and so on leading to an infinite regress. Of course this anthropomorphic manner of speaking about the formation of sets is only that; a manner of speaking. We humans do not actually physically form sets out of their elements, as sets are abstract objects. Nevertheless the sets that have been needed to represent the standard abstract objects of modern mathematics have, in fact, been the ordinary well-founded ones. This observation has been institutionalised in the standard axiom system ZFC of axiomatic set theory, which includes among its axioms the foundation axiom FA. This axiom simply expresses that all sets are well-founded.

If non-well-founded sets are not needed for the development of mathematics then it may well seem natural to leave them out of consideration when formulating an axiomatic basis for mathematics. Sometimes a stronger view is expressed. According to that view there is only one sensible coherent notion of set. That is the iterative conception in which sets are arranged in levels, with the elements of a set placed at lower levels than the set itself. For the iterative conception only well-founded sets exist and FA and the other axioms of ZFC are true when interpreted in the iterative universe of pure sets. There has been yet one more reason

---

* See the epigraph.

why *FA* has been routinely included among the axioms of axiomatic set theory. This is the fact that the cumulative hierarchy of the iterative universe has an enticingly elegant mathematical structure. This structure was already revealed by Mirimanoff and over the years it has been powerfully exploited by set theorists in a great variety of model constructions. There is a natural reluctance to forgo the pleasure of working within this structure. Certainly I myself must admit to having been rather seduced by it. But there have been doubts about the coherence of the iterative conception. Part of its appeal has been the essentially naive but very intuitive image of a set being physically formed out of its elements. This image is translated to an abstract realm and given some plausibility by a sometimes subconcious suggestion of constructivity. The suggestion is to take the abstract realm to be a realm of mental constructions. In fact such a suggestion cannot easily be sustained (not by me anyway) and one is forced to a Platonistic conception in which sets are taken to have a non-physical existence independent of us.

The purpose of Part One of this book is to investigate the axiom system obtained by replacing *FA* in *ZFC* by an axiom that I have chosen to call the Anti-Foundation Axiom, abbreviated *AFA*. This axiom expresses, in a particular way, that every possible non-well-founded set exists. The resulting axiom system is $ZFC^- + AFA$, where ZFC^- is *ZFC* without *FA*.

There are other variants to *AFA* which also can be viewed as expressing that every possible non-well-founded set exists. I call these axioms collectively anti-foundation axioms, reserving *AFA* for *the* anti-foundation axiom. Some of these variants are discussed in Part Two. The reason for the existence of more than one anti-foundation axiom is the fact that there is more than one criterion for equality between sets when non-well-founded sets are allowed. One approach is to keep to the extensionality criterion of *ZFC* as the sole one, even for possibly non-well-founded sets; so sets are equal if they have the same elements and nothing further is to be stipulated concerning set equality. This is the approach that was developed in a series of papers in the 1960s and early 1970s by Maurice Boffa and is here presented in chapter 5, where we call the resulting anti-foundation axiom *BAFA*. The other variants of *AFA* do use a strengthening of extensionality as a criterion for set equality. It turns out that these other approaches can be treated in a uniform manner and this leads to the formulation of an axiom $AFA^\sim$ relative to the definition of a suitable relation $\sim$ to express the criterion of set equality. This is presented in Chapter 4. The suitable relations $\sim$ are called regular bisimulation relations and range between two possible extremes. One extreme is the maximal bisimulation relation on the universe of sets. This relation gives the most generous criterion for set equality which roughly states that sets are equal whenever possible, keeping in mind that if two sets are equal then any element of one set must be equal to an element of the other set. It is this relation that gives rise to the axiom AFA. There is the other extreme of a strengthening of the extensionality criterion for set equality which roughly states that two sets are equal if they are isomorphic in a suitable sense. This gives rise to an anti-foundation axiom that we call $FAFA$. It turns out that there is an alternative notion of isomorphism between sets which gives rise to yet another anti-foundation axiom which we call $SAFA$. In all we consider the four specific anti-foundation axioms AFA, $BAFA$, $FAFA$ and $SAFA$. Each can be consistently added to the axiom system ZFC^- and each gives rise to an axiom system in which every possible non-well-founded set exists when account is taken of the particular criterion of set equality that is being used. Nevertheless the four axiom systems are incomparable in the sense that in ZFC^- no one of the four axioms AFA, $SAFA$, $FAFA$ or $BAFA$ can be proved from any other, assuming that ZFC^- is consistent.

Each of the four anti-foundation axioms was first formulated in one way or another by someone else. Nevertheless here I attempt to consider them all in a uniform setting. So I have chosen to introduce my own more uniform terminology. The reader should examine the Notes towards a History, at the end of the book, to find out out more about the earlier work.

The original stimulus for my own interest in the notion of a non-well-founded set came from a reading of the work of Robin Milner in connection with his development of a mathematical theory of concurrent processes. This topic in theoretical computer science is one of a number of such topics that are generating exciting new ideas and intuitions that are in need of suitable mathematical expression. In chapter 8 I outline how I see the relationship between Milner's ideas and the axiom AFA and non-well-founded sets.

Another major area of application for the notion of a non-well-founded set and the axiom AFA is to situation theory. Jon Barwise realised the significance of AFA for situation theory while I was giving the lectures that form the origin of this book. I have chosen not to present any of the details of this application here. Instead, in chapters 6 and 7, I have focussed attention on what I consider to be some of the fundamental general mathematical ideas that are being exploited when using AFA. Some of the ideas and terminology have been presented in an elegant and appealing way in the book *The Liar*, by Jon Barwise and John Etchemendy, and I have taken the opportunity to incorporate those ideas into this book.

Part One

The Anti-Foundation Axiom

1 | Introducing the Axiom

Pictures Of Sets

Sets may be pictured using (downward growing) trees. For example if we use the standard set theoretical representation of the natural numbers, where the natural number n is represented as the set of natural numbers less than n, then we have the following pictures for the first few natural numbers:

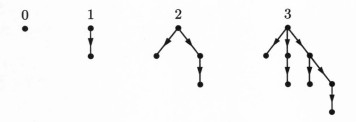

More generally pointed graphs may be used as pictures of sets. For example we have the following alternative pictures of 2 and 3:

So what exactly is a picture of a set? We need some terminology. Here a GRAPH will consist of a set of NODES and a set

3

of EDGES, each edge being an ordered pair (n, n') of nodes. If (n, n') is an edge then I will write $n \longrightarrow n'$ and say that n' is a CHILD of n. A PATH is a finite or infinite sequence

$$n_0 \longrightarrow n_1 \longrightarrow n_2 \longrightarrow \ldots$$

of nodes n_0, n_1, n_2, ... linked by edges (n_0, n_1), (n_1, n_2), ... A POINTED GRAPH is a graph together with a distinguished node called its POINT. A pointed graph is ACCESSIBLE if for every node n there is a path $n_0 \longrightarrow n_1 \longrightarrow \cdots \longrightarrow n$ from the point n_0 to the node n. If this path is always unique then the pointed graph is a TREE and the point is the ROOT of the tree. We will use accessible pointed graphs (apgs for short) as our pictures. In the diagrams the point will always be located at the top. A DECORATION of a graph is an assignment of a set to each node of the graph in such a way that the elements of the set assigned to a node are the sets assigned to the children of that node. A PICTURE of a set is an apg which has a decoration in which the set is assigned to the point.

Notice that in our examples there is only one way to decorate the apgs. For example the last diagram must be decorated in the following way.

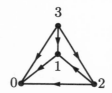

The node labelled 0 has no children and hence must be assigned the empty set, i.e. 0, in any decoration. The central node has as only child the node labelled 0. Hence in any decoration the central node must be assigned the set $\{0\}$, i.e. 1. Continuing in this way we are inevitably led to the above decoration, and this decoration shows us that we have a picture of 3. Reflecting on how the decoration was formed we are led to a formulation of an important result of set theory. Call a graph WELL-FOUNDED if it has no infinite path.

Mostowski's Collapsing Lemma:
Every well-founded graph has a unique decoration.

This result is proved by a simple application of definition by recursion on a well-founded relation to obtain the unique function d defined so that

$$dn = \{dn' \mid n \longrightarrow n'\}$$

for each node n of the graph. The decoration d assigns the set dn to the node n. Note the obvious consequence.

1.1 Corollary:
Every well-founded apg is a picture of a unique set.

Which sets have pictures? There is a simple answer to this question.

1.2 Proposition: *Every set has a picture.*

To see this we will associate with each set a its CANONICAL PICTURE. Form the graph that has as its nodes those sets that occur in sequences $a_0, a_1, a_2, \ldots$ such that

$$\ldots \in a_2 \in a_1 \in a_0 = a$$

and having as edges those pairs of nodes (x, y) such that $y \in x$. If a is chosen as the point we obtain an apg. This apg is clearly a picture of a, the decoration consisting of the assignment of the set x to each node x. Note that this construction does not require the set a to be well-founded.

Every picture of a set can be unfolded into a tree picture of the same set. Given an apg we may form the tree whose nodes are the finite paths of the apg that start from the point of the apg and whose edges are pairs of paths of the form

$$(a_0 \longrightarrow \cdots \longrightarrow a, \ a_0 \longrightarrow \cdots \longrightarrow a \longrightarrow a').$$

The root of this tree is the path a_0 of length one. This tree is the UNFOLDING of the apg. Any decoration of the apg induces a decoration of its unfolding by assigning to the node $a_0 \longrightarrow \cdots \longrightarrow a$ of the tree the set that is assigned to the node a of the apg by the decoration of the apg. Thus the unfolding of an apg will picture any set pictured by the apg. The unfolding of the canonical picture of a set will be called the CANONICAL TREE PICTURE of the set.

Our discussion so far has been intended to motivate the following axiom:

The Anti-Foundation Axiom, AFA:

Every graph has a unique decoration.

Note the following obvious consequences.

- Every apg is a picture of a unique set.
- Non-well-founded sets exist.

In fact any non-well-founded apg will have to picture a non-well-founded set.

Examples of Non-Well-Founded Sets

In the rest of this section we will examine some pictures of non-well-founded sets assuming the new axiom. Of course we must relinquish the foundation axiom, but it will turn out that we need drop none of the other axioms of set theory.

1.3 Example: *Consider the apg*

This is a picture of the unique set Ω such that

$$\Omega = \{\Omega\}.$$

This is our first example of a non-well-founded set. When the apg above is unfolded we get the infinite tree

An analogous 'unfolding' of the equation above would seem to give us

$$\Omega = \{\{\{\cdots\}\}\},$$

if only the infinite expression on the right hand side had an independently determined meaning!

The infinite tree above and the infinite expression associated with it might suggest that in some sense Ω is an infinite object. But a moment's thought should convince the reader that Ω is as finite an object as one could wish. After all it does have a finite picture. We may call sets that have finite pictures HEREDITARILY FINITE sets.

Ω has many pictures. In fact we have the following characterisation.

1.4 Proposition: *An apg is a picture of Ω if and only if every node of the apg has a child.*

Proof: Assume given a picture of Ω with root a. Let d be a decoration of the picture such that $da = \Omega$. Now if b is any node of the picture there must be a path $a = a_0 \longrightarrow \cdots \longrightarrow a_n = b$ so that $db = da_n \in \cdots \in da_0 = da = \Omega$. As Ω is the only element of Ω it follows that $db = \Omega$. As Ω has an element it follows that b must have a child. Thus every node of the picture must have a child.

Conversely assume given an apg with the property that every node has a child. Then the assignment of Ω to each node of the apg is easily seen to be a decoration of the apg, so that the apg is a picture of Ω. □

1.5 Example: *The apg*

is a picture of the unique set 0^* such that

$$0^* = \{0, 0^*\}.$$

When "unfolded" this equation becomes

$$0^* = \{0, \{0, \{0, \ldots\}\}\}.$$

1.6 Example: *We have seen that every set has a picture. Let*

denote a picture of some set a. Then

is a picture of the unique set a^* such that

$$a^* = \{a, a^*\}.$$

If $a = 0$ we get the special case in example 1.5. Again the above equation can be 'unfolded' in the obvious way.

Let us now consider the special case when $a = \Omega$. Ω^* is the unique set such that $\Omega^* = \{\Omega, \Omega^*\}$. But $\Omega = \{\Omega\} = \{\Omega, \Omega\}$. Hence we must conclude that $\Omega^* = \Omega$. Of course this is also clear from the characterisation of pictures of Ω given earlier.

1.7 Example: *The ordered pair of two sets is usually represented as follows:*

$$(a, b) = \{\{a\}, \{a, b\}\}.$$

So the equation

$$x = (0, x)$$

becomes

$$x = \{\{0\}, \{0, x\}\}.$$

This equation in one variable x is equivalent in an obvious sense, to the following system of four equations in the four variables x, y, z, w.

$$x = \{y, z\}$$
$$y = \{w\}$$
$$z = \{w, x\}$$
$$w = 0$$

Now these equations hold exactly when the following diagram is of a correctly decorated apg.

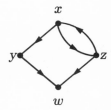

Hence by *AFA* the above system of four equations has a unique solution and hence the original equation

$$x = (0, x)$$

has a unique solution with picture

"Unfolding" this equation we get

$$x = (0, (0, (0, \ldots)))$$

1.8 Example: *As in example 1.6 the previous example may be generalised to show that for any set a the equation $x = (a, x)$ has a unique solution $x = (a, (a, (a, \ldots)))$.*

More generally still, given any infinite sequence of sets a_0, a_1, a_2, ... we may consider the following infinite system of equations *in the variables x_0, x_1, x_2, x_3, ...*

$$x_0 = (a_0, x_1)$$
$$x_1 = (a_1, x_2)$$
$$x_2 = (a_2, x_3)$$

.

.

.

It should be a straightforward exercise for the reader to show that this system of equations has a unique solution. Infinite expressions for this unique solution can be obtained by 'unfolding' the system of equations to get

$$x_0 = (a_0, (a_1, (a_2, \ldots)))$$
$$x_1 = (a_1, (a_2, (a_3, \ldots)))$$
$$x_2 = (a_2, (a_3, (a_4, \ldots)))$$

.

.

.

This and other examples can be treated even more simply using the following strengthening of *AFA*. A LABELLED GRAPH is a graph together with an assignment of a set $a{\downarrow}$ of LABELS to each node a.

A LABELLED DECORATION of a labelled graph is an assignment d of a set da to each node a such that

$$da = \{db \mid a \rightarrow b\} \cup a{\downarrow} \, .$$

The Labelled Anti-Foundation Axiom:
   Every labelled graph has a unique labelled decoration.

The ordinary anti-foundation axiom may be viewed as a special case by treating ordinary graphs as labelled graphs with an empty set of labels attached to each node. Conversely, we will see that the labelled anti-foundation axiom is a consequence of the ordinary one.

Now given sets $a_0, a_1, a_2, \ldots$ we can obtain the sets x_n such that $x_n = (a_n, x_{n+1})$ for $n = 0, 1, \ldots$ in the following way. Consider the labelled graph having natural numbers as nodes, an edge $n \to n+1$ for each n and sets of labels given by:

$$(2n)\!\downarrow = \{\{a_n\}\},$$
$$(2n+1)\!\downarrow = \{a_n\}.$$

Using the labelled anti-foundation axiom let d be the unique labelled decoration of the labelled graph.

Then for $n = 0, 1, \ldots$

$$d(2n) = \{d(2n+1)\} \cup \{\{a_n\}\},$$
$$d(2n+1) = \{d(2n+2)\} \cup \{a_n\}.$$

Hence if $x_n = d(2n)$ then

$$\begin{aligned} x_n &= \{d(2n+1), \{a_n\}\} \\ &= \{\{x_{n+1}, a_n\}, \{a_n\}\} \\ &= (a_n, x_{n+1}) \end{aligned}$$

for each n. Hence we have obtained the desired sets x_n. Their uniqueness easily follows from the uniqueness of d.

There is an even more powerful technique that can be used to deal with this and other examples. This technique involves the formulation of a result asserting that every system of equations of a certain type has a unique solution. We can then simply apply the result directly to each example without the need for any coding. Following the terminology of Barwise and Etchemendy the result will be called the solution lemma below. In order to formulate the lemma in an intuitively appealing way we need to consider an expansion of the universe of pure sets that we have been considering so far. Pure sets can only have sets as elements and those sets are also pure. The expansion of the universe involves the addition of atoms and sets built out of them. Atoms are objects that are not sets and are not made up of sets in any way, so that they have no set theoretical structure. But they can be used in the formation of sets. See (Barwise 1975) for a discussion of the formalisation of set theory with atoms. In that book atoms are called Urelemente. The construction of an expanded universe by adjoining to a universe of sets atoms and adding all the sets that can involve these atoms in their build up is analogous to the construction of a polynomial ring from a ring by adjoining indeterminates and adding all the polynomials in those indeterminates with coefficients taken from the ring. It will be convenient to assume that we have a plentiful supply of atoms. So we assume that for each pure set i there is an atom x_i, with $x_i \neq x_j$ for distinct pure sets i, j. If X is a class of atoms then we call sets that may involve atoms from the class X in their build up X-SETS. The solution lemma will apply to a system of equations of the form

$$x = a_x \quad (x \in X),$$

where a_x is an X-set for each $x \in X$. For example given pure sets $a_0, a_1, \ldots$ the system of equations

$$x_n = (a_n, x_{n+1}) \quad (n = 0, 1, \ldots)$$

has the above form when we take $X = \{x_0, x_1, \ldots\}$ and for each n we take a_{x_n} to be (a_n, x_{n+1}); i.e. the X-set $\{\{a_n\}, \{a_n, x_{n+1}\}\}$. In this example it is clear what a solution of this system of equations must be. It is a family of pure sets $b_0, b_1, \ldots$, one for each atom in X, such that

$$b_n = (a_n, b_{n+1}) \text{ for } n = 0, 1, \ldots.$$

Notice that the right hand sides of these equations are obtained from the right hand sides of the original system of equations by substituting b_n for each atom x_n. This suggests what a solution to the general system of equations should be. It should be a family $\pi = (b_x)_{x \in X}$ of pure sets b_x, one for each $x \in X$, such that for each $x \in X$

$$b_x = \hat{\pi} a_x.$$

Here, for each X-set a, the set $\hat{\pi} a$ is that pure set that is obtained from a by substituting b_x for each occurence of an atom x in the build up of a. So $\hat{\pi}$ is the substitution operation characterised in the following result.

Substitution Lemma:
For each family of pure sets $\pi = (b_x)_{x \in X}$ there is a unique operation $\hat{\pi}$ that assigns a pure set $\hat{\pi} a$ to each X-set a such that

$$\hat{\pi} a = \{\hat{\pi} b \mid b \text{ is an } X\text{-set such that } b \in a\} \cup \{\pi x \mid x \in a \cap X\}.$$

We can now state the result we have been aiming at.

Solution Lemma:
If a_x is an X-set for each atom x in the class X of atoms then the system of equations

$$x = a_x \quad (x \in X)$$

has a unique solution; i.e. a unique family of pure sets $\pi = (b_x)_{x \in X}$ such that for each $x \in X$

$$b_x = \hat{\pi} a_x.$$

The above informal discusion of the solution lemma seems to be all that is required when seeking to apply the lemma. A rigorous formulation and proof will be left till the end of this chapter.

Systems

We need to widen the notion of a graph so as to allow there to be a proper class of nodes. A SYSTEM is a class M of nodes together with a class of EDGES consisting of ordered pairs of nodes. We shall simply use M to refer to the system and write that $a \to b$ in M or simply $a \to b$ if (a, b) is an edge of M. A system M is required to satisfy the condition that for each node a the class $a_M = \{b \in M \mid a \to b\}$ of children of a is a set.

Note that a graph is simply a small system. An example of a large system is the universe V with $a \to b$ whenever $b \in a$.

The notion of a decoration of a graph extends to systems in the obvious way. We get the following strengthening of *AFA*.

1.9 Theorem: *(assuming AFA)*
    Each system has a unique decoration.

Proof: Let M be a system. To each $a \in M$ we may associate an apg Ma constructed as follows:

- The nodes and edges of Ma are those nodes and edges of M that lie on paths of M starting from the node a, and the point of Ma is the node a itself.

That the nodes of Ma do form a set may be seen as follows. Let $X_0 = \{a\}$ and for each natural number n let

$$X_{n+1} = \bigcup \{x_M \mid x \in X_n\}.$$

As x_M is a set for all $x \in M$, each X_n is a set, the set of those nodes of M that end paths in M of length n starting from the node a. So the nodes of Ma form the set $\bigcup_n X_n$.

By AFA each apg Ma has a unique decoration d_a so that Ma will be a picture of the set $d_a a$. For each $a \in M$ let $da = d_a a$. We will show that d is the unique decoration of M. First observe that if $a \to x$ in M then every node of Mx will also be a node of Ma and the restriction of d_a to Mx will be a decoration of Mx and hence equal to d_x, the uniquè decoration of Mx. Hence if $a \to x$ in M then $d_a x = d_x x = dx$.

So, for each $a \in M$,

$$
\begin{aligned}
da &= d_a a \\
&= \{d_a x \mid a \to x \text{ in } Ma\} \\
&= \{dx \mid a \to x \text{ in } M\}.
\end{aligned}
$$

Thus d is a decoration of M. To see the uniqueness of this decoration it suffices to observe that any decoration of M must be a decoration of each Ma, when restricted, and hence must extend each d_a, so that it must be d itself. □

LABELLED SYSTEMS and their labelled decorations are defined in the obvious way. If $a \in M$ then $a{\downarrow}M$ will denote the set of labels at a in the labelled system M. We next generalise the previous result to labelled systems.

1.10 Theorem: *(assuming AFA)*
Each labelled system has a unique labelled decoration.

Proof: Let M be a labelled system. Let M' be the system having as nodes all the ordered pairs (i, a) such that either $i = 1$ and $a \in M$ or $i = 2$ and $a \in V$ and having as edges:

- $(1, a) \to (1, b)$ whenever $a \to b$ in M,
- $(1, a) \to (2, b)$ whenever $a \in M$ and $b \in a{\downarrow}M$,
- $(2, a) \to (2, b)$ whenever $b \in a$.

By AFA M' has a unique decoration π. So for each $a \in M$

$$\pi(1, a) = \{\pi(1, b) \mid a \to b \text{ in } M\} \cup \{\pi(2, b) \mid b \in a{\downarrow}M\}$$

and for each $a \in V$

$$\pi(2, a) = \{\pi(2, b) \mid b \in a\}.$$

Note that the assigment of the set $\pi(2, a)$ to each $a \in V$ is a decoration of the system V so that by AFA $\pi(2, a) = a$ for all $a \in V$. Hence if we let $\tau a = \pi(1, a)$ for $a \in M$ then, for $a \in M$,

$$\tau a = \{\tau b \mid a \to b \text{ in } M\} \cup a{\downarrow}M,$$

so that τ is a labelled decoration of the labelled system M.

For the uniqueness of τ suppose that τ' is a labelled decoration of the labelled system M. Then π' is a decoration of the system M' where

$$\pi'(1, a) = \tau'a \quad \text{for } a \in M,$$
$$\pi'(2, a) = a \quad\;\; \text{for } a \in V.$$

It follows from AFA that $\pi' = \pi$ so that for $a \in M$

$$\tau'a = \pi'(1, a) = \pi(1, a) = \tau a,$$

and hence $\tau' = \tau$. $\qquad\qquad\qquad\qquad\qquad\qquad\qquad\qquad$ □

We next give a general result that will then be used to prove the Substitution and Solution Lemmas.

1.11 Theorem: *(assuming AFA) Let M be a labelled system whose sets of labels are subsets of the class X.*

(1) *If $\pi : X \to V$ then there is a unique map $\hat{\pi} : M \to V$ such that for each $a \in M$*

$$\hat{\pi} a = \{\hat{\pi} b \mid a \to b \text{ in } M\} \cup \{\pi x \mid x \in a{\downarrow}M\}.$$

(2) *Given $a_x \in M$ for $x \in X$ there is a unique map $\pi : X \to V$ such that for all $x \in X$*

$$\pi x = \hat{\pi} a_x.$$

Proof:

(1) For each $\pi : X \to V$ let M_π be the labelled system obtained from M by redefining the sets of labels so that for each node a
$$a{\downarrow}M_\pi = \{\pi x \mid x \in a{\downarrow}M\}.$$

Then the required unique map $\hat{\pi}$ is the unique labelled decoration of M_π.

(2) Let M' be the system having the same nodes as M, and all the edges of M together with edges $a \to a_x$ whenever $a \in M$ and $x \in a{\downarrow}M$. By theorem 1.9 M' has a unique decoration φ. So for each $a \in M$

$$\varphi a = \{\varphi b \mid a \to b \text{ in } M\} \cup \{\varphi a_x \mid x \in a{\downarrow}M\}.$$

Let $\pi x = \varphi a_x$ for $x \in X$. Then φ is a labelled decoration of the labelled system M_π so that $\varphi = \hat\pi$ and hence $\pi x = \varphi a_x$ for $x \in X$. For the uniqueness of π let $\pi' : X \to V$ such that $\pi'x = \hat\pi'a_x$ for $x \in X$. Then observe that $\hat\pi'$ is a decoration of M', so that $\hat\pi' = \varphi$ and hence $\pi'x = \hat\pi'a_x = \varphi a_x = \pi x$ for $x \in X$. So $\pi' = \pi$.   □

Proof of the Substitution and Solution Lemmas

The informal presentation of the substitution and solution lemmas that we have given cannot be made rigorous in a direct way on the basis of the axiom system for set theory that we have been implicitly working in. Rather than modify this axiom system, so as to be appropriate for the expanded universe with atoms, we will give a model of the expanded universe within the universe of pure sets. We will use the pure sets $x_i = (1, i)$ to be the atoms in the model and will call them *-atoms. The sets in the model will be called *-sets and will be certain pure sets of the form $(2, u)$. If $a = (2, u)$ then let $a^* = u$. The elements of a^* will be called the *-elements of a. The class of *-sets is defined to be the largest class of sets of the form $(2, u)$ such that each *-element of a *-set is either a *-atom or else is a *-set. We will not stop here to show the existence of such a largest class, but refer the reader to theorem 6.5. Given a class X of *-atoms we also define the class of X-sets to be the largest class of *-sets such that every *-atom in an X-set is in X. Now the X-sets form the class of nodes of a labelled system M, where for each node a

$$\begin{aligned} a_M &= M \cap a^*, \\ a \downarrow M &= X \cap a^*. \end{aligned}$$

We may apply the two parts of theorem 1.11 to this labelled system to obtain proofs of the substitution and solution lemmas, except that in the right hand side of the characterising equation for $\hat\pi a$ in the substitution lemma, the set a must be replaced by the set a^*. This slight revision of the substitution lemma is needed as we are not really expanding the universe but only using a model of an expanded universe.

2 | The Axiom in More Detail

The anti-foundation axiom is obviously equivalent to the conjunction of the following two statements

- AFA_1: Every graph has at least one decoration.
- AFA_2: Every graph has at most one decoration.

In this chapter we shall give equivalent formulations of each of these statements. An equivalent of AFA_2 will make clear that it expresses a criterion of equality for possibly non-well-founded sets. Other equivalents for AFA_1 and AFA_2 will yield an equivalent for AFA which expresses an answer to the question

Which apgs are isomorphic to canonical pictures?

The chapter will end by considering an interesting consequence *NSA* of AFA_1.

Let us consider the question of set equality. We are familiar with the extensionality criterion that two sets are equal if they have the same elements. Can there be more to say about equality between sets? For well-founded sets the answer is no. This is because as soon as the equality relation between the elements of two sets has been fixed, the extensionality criterion determines the conditions of equality for the two sets. So by a transfinite induction on the membership relation the equality relation between well-founded sets is uniquely determined. But now consider the equation

$$x = \{x\}.$$

Can there be distinct sets that satisfy this equation? The extensionality axiom does not help us to answer this question. While AFA implies that there is at most one solution to the equation, namely Ω, it is in fact consistent to suppose that there are many solutions. This example shows that if one is to attempt to formulate a sensible notion of non-well-founded set it is worthwhile to strengthen the extensionality axiom.

2.1 Exercise: *Show that the foundation axiom implies AFA_2 and the negation of AFA_1.*

For sets a, b let $a \equiv b$ if and only if there is an apg that is a picture of both a and b.

2.2 Exercise: *Show that AFA_2 is equivalent to:*

$$a \equiv b \implies a = b \quad \text{for all sets } a, b.$$

Bisimulations

What sort of relation is $\equiv$? To start to answer this question we need the following fundamental notion. A binary relation R on the system M is a BISIMULATION on M if $R \subseteq R^+$, where for $a, b \in M$

$$aR^+b \iff \forall x \in a_M \exists y \in b_M \ xRy \quad \& \quad \forall y \in b_M \exists x \in a_M \ xRy.$$

Observe that if $R_0 \subseteq R$ then $R_0^+ \subseteq R^+$; i.e. the operation $(\ )^+$ is monotone.

2.3 Exercise: *Show that the relation $\equiv$ is a bisimulation on V.*

In general a system M will have many bisimulations. We will see that $\equiv$ is the maximum bisimulation on the system V. A maximum bisimulation exists on any system.

2.4 Theorem: *There is a unique maximum bisimulation $\equiv_M$ on each system M; i.e.*

(1) $\equiv_M$ *is a bisimulation on M,*

(2) *If R is a bisimulation on M then for all $a, b \in M$*

$$aRb \implies a \equiv_M b.$$

In fact

$$a \equiv_M b \iff aRb \quad \text{for some small bisimulation } R \text{ on } M.$$

The relation $\equiv_M$ is also sometimes called the weakest bisimulation or largest bisimulation on M.

Proof: Let $\equiv_M$ be defined as above. We prove (1) and (2). For (1), let $a \equiv_M b$. Then aRb for some small bisimulation R on M. Note that, trivially,

$$xRy \implies x \equiv_M y \quad \text{for all } x, y \in M,$$

so that by the monotonicity of $(\ )^+$

$$xR^+y \implies x \equiv_M^+ y \quad \text{for all } x, y \in M.$$

As aRb and $R \subseteq R^+$ it follows that $a \equiv_M^+ b$. For (2) let R be a bisimulation on M and aRb. Recall from the proof of theorem 1.9 that for each $x \in M$ Mx is an apg with point x such that for $u, v \in Mx$

$$u \to v \text{ in } Mx \iff u \to v \text{ in } M.$$

It is easy to check that if

$$R_0 = R \cap ((Ma) \times (Mb))$$

then R_0 is a bisimulation on M such that aR_0b. Moreover, as Ma and Mb are small the bisimulation R_0 is small. So $a \equiv_M b$.
□

2.5 Proposition: *For all sets a, b*

$$a \equiv b \iff a \equiv_V b.$$

Proof: As $\equiv$ is a bisimulation on V (by exercise 2.3), and $\equiv_V$ is the maximum bisimulation on V it follows that the implication from left to right holds. For the converse implication it suffices to show that if R is a bisimulation on V then for all sets a, b

$$aRb \implies a \equiv b.$$

So let R be a bisimulation on V. Define the system M_0 as follows. The nodes of M_0 are the elements of R; i.e. the ordered pairs (a, b) such that aRb. The edges of M_0 are defined so that

$$(a, b) \to (x, y) \text{ in } M_0 \iff x \in a \quad \& \quad y \in b.$$

Now observe that d_1 and d_2 are both decorations of M_0, where for $(a, b) \in M_0$

$$d_1(a, b) = a,$$
$$d_2(a, b) = b.$$

So, if aRb then the apg $M_0(a, b)$ is a picture of both a and b, using the restrictions of d_1 and d_2 to the apg. Hence if aRb then $a \equiv b$. $\qquad \square$

The facts in the following exercise will be useful in showing that the maximum bisimulation relation on a system is an equivalence relation.

2.6 Exercise: *Show that if M is a system then*

(i) *For all $a, b \in M$*

$$a =_M^+ b \iff a_M = b_M,$$

(ii) *If $R \subseteq M \times M$ then*

$$(R^{-1})^+ = (R^+)^{-1},$$

(iii) *If $R_1, R_2 \subseteq M \times M$ then*

$$R_1^+ \mid R_2^+ \subseteq (R_1 \mid R_2)^+.$$

2.7 Proposition: *For each system M the relation $\equiv_M$ is an equivalence relation on M such that for all $a, b \in M$*

$$a \equiv_M^+ b \iff a \equiv_M b.$$

Proof: That $\equiv_M$ is an equivalence relation is an easy application of the previous exercise. As $\equiv_M$ is a bisimulation we have the implication from right to left. As the operation $(\ )^+$ is monotone it follows that $\equiv_M^+$ is also a bisimulation. As $\equiv_M$ is the maximum bisimulation we get the implication from left to right. $\qquad \square$

2.8 Exercise: *If M is a system show that for all $a, b \in M$*

(i) $a_M = b_M \implies a \equiv_M b,$

(ii) $Ma \cong Mb \implies a \equiv_M b.$

In (ii) Ma is the apg determined from M and a in the proof of theorem 1.9, and $\cong$ is the isomorphism relation between apgs.

A system M is EXTENSIONAL if, for all $a, b \in M$

$$a_M = b_M \implies a = b.$$

It is STRONGLY EXTENSIONAL if, for all $a, b \in M$

$$a \equiv_M b \implies a = b.$$

2.9 Exercise: *Show that*

$$AFA_2 \iff AFA_2^{ext},$$

where AFA_2^{ext} *is:*

Every extensional graph has at most one decoration.

Observe that by (i) of exercise 2.8 every strongly extensional system is extensional. Note that by the extensionality axiom the system V is extensional. By the next result AFA_2 expresses a strengthening of the extensionality axiom.

2.10 Proposition: $AFA_2 \iff V$ *is strongly extensional.*

Proof: Let us first assume AFA_2 and let $a \equiv_V b$. Then by exercise 2.5 $a \equiv b$ so that there is an apg Gn and decorations d_1 and d_2 of G such that $d_1 n = a$ and $d_2 n = b$. By AFA_2 $d_1 = d_2$ so that $a = b$. Thus V is strongly extensional.

Conversely let V be strongly extensional and let d_1 and d_2 be decorations of a graph G. If $x \in G$ then $d_1 x \equiv d_2 x$, as Gx is a picture of both $d_1 x$ and $d_2 x$. Hence, by exercise 2.5 $d_1 x \equiv_V d_2 x$, so that $d_1 x = d_2 x$, as V is strongly extensional. Thus $d_1 = d_2$ so that we have proved AFA_2. $\qquad\square$

System Maps

A SYSTEM MAP from the system M to the system M' is a map $\pi : M \longrightarrow M'$ such that for $a \in M$

$$(\pi a)_{M'} = \{\pi b \mid b \in a_M\}.$$

If π is a bijection then it is a SYSTEM ISOMORPHISM.

2.11 Example: *A system map* $G \longrightarrow V$, *where* G *is a graph, is simply a decoration of the graph.*

2.12 Exercise: *Show that systems and system maps form a (superlarge) category.*

The close relationship which exists between bisimulations and system maps is illustrated by the following results.

2.13 Proposition: *Let $\pi_1, \pi_2 : M \longrightarrow M'$ be system maps.*

(1) *If R is a bisimulation on M then*

$$(\pi_1 \times \pi_2)R \overset{\text{def}}{=} \{(\pi_1 a_1, \pi_2 a_2) \mid a_1 R a_2\}$$

is a bisimulation on M'.

(2) *If S is a bisimulation on M' then*

$$(\pi_1 \times \pi_2)^{-1}S \overset{\text{def}}{=} \{(a_1, a_2) \in M \times M \mid (\pi_1 a_1)S(\pi_2 a_2)\}$$

is a bisimulation on M.

Proof:

(1) Let $S = (\pi_1 \times \pi_2)R$ and let $b_1 S b_2$ and $b_1' \in b_{1M'}$. Then there are a_1, a_2 such that $a_1 R a_2$ and $b_1 = \pi_1 a_1$, $b_2 = \pi_2 a_2$. As $b_1' \in (\pi a_1)_{M'}$ there is $a_1' \in (a_1)_{M'}$ such that $b_1' = \pi a_1'$. As R is a bisimulation on M there is $a_2' \in (a_2)_M$ such that $a_1' R a_2'$. Now if $b_2' = \pi a_2'$ then $b_1' S b_2'$ and $b_2' \in (b_2)_{M'}$. Thus we have proved that if $b_1 S b_2$ then

$$\forall b_1' \in (b_1)_{M'} \; \exists b_2' \in (b_2)_{M'} \; b_1' S b_2'.$$

Similarly, if $b_1 S b_2$ then also

$$\forall b_2' \in (b_2)_{M'} \; \exists b_1' \in (b_1)_{M'} \; b_1' S b_2'.$$

Thus S is a bisimulation on M'.

(2) Let $R = (\pi_1 \times \pi_2)^{-1}S$ and let $a_1 R a_2$ and $a_1' \in a_M$. Then $(\pi_1 a_1)S(\pi_2 a_2)$ and $\pi_1 a_1' \in (\pi_1 a_1)_{M'}$. As S is a bisimulation on M' there is $b_2' \in (\pi_2 a_2)_{M'}$ such that $(\pi_1 a_1')S b_2'$. So $b_2' = \pi_2 a_2'$ for some $a_2' \in (a_2)_M$. So $a_1' R a_2'$ for some $a_2' \in (a_2)_M$. Thus if $a_1 R a_2$ then

$$\forall a_1' \in (a_1)_M \; \exists a_2' \in (a_2)_M \; a_1' R a_2',$$

and similarly we get

$$\forall a_2' \in (a_2)_M \; \exists a_1' \in (a_1)_M \; a_1' R a_2'.$$

$\square$

2.14 Exercise: *Let R be a bisimulation on M. Let M_0 be the system whose nodes are the ordered pairs in the relation R with*

$$(a,b) \;\to\; (a',b') \text{ in } M_0 \quad \text{iff } a \,\to\, a' \text{ and } b \,\to\, b' \text{ in } M.$$

Let $\pi_1, \pi_2 : M_0 \to M$ be given by

$$\pi_1(a,b) = a,$$
$$\pi_2(a,b) = b,$$

for $a,b \in M$.
Show that π_1 and π_2 are system maps.

This exercise generalises a construction in the proof of proposition 2.5 to an arbitrary system M.

2.15 Exercise: *If M is a system show that $a \equiv_M b$ if and only if there is a graph G and system maps $d_1, d_2 : G \to M$ such that $a = d_1 n$ and $b = d_2 n$ for some $n \in G$.*

This exercise generalises proposition 2.5 and was suggested by Dag Westerståhl.

Let $\pi : M \to M'$ be a quotient of the class M with respect to the equivalence relation R on M; i.e. π is a surjective map such that for all $a,b \in M$

$$aRb \iff \pi a = \pi b.$$

Now suppose that M is a system and R is a bisimulation on M. Then π is a system map, provided that M' is made into a system by letting the edges of M' be all the pairs $(\pi a, \pi b)$ for (a,b) an edge of M. We will say that $\pi : M \to M'$ (or sometimes, simply M') is a QUOTIENT of the system M with respect to the bisimulation equivalence R. Note that any two such quotients will be isomorphic systems.

2.16 Exercise: *Show that if $\pi : M \to M'$ is a quotient of the system M with respect to a bisimulation equivalence R then M' is strongly extensional if and only if R is the relation $\equiv_M$.*

If $\pi : M \to M'$ is a quotient of the system M with respect to $\equiv_M$ then we say that it is a STRONGLY EXTENSIONAL QUOTIENT of M.

2.17 Lemma: *Every system has a strongly extensional quotient.*

Proof: The essential problem is to define a map π with domain the system M such that for $a_1, a_2 \in M$

$$a_1 \equiv_M a_2 \iff \pi a_1 = \pi a_2.$$

For small M the standard definition of π in terms of equivalence classes would work. In general a strong form of global choice would be needed to pick a representative from each equivalence class. Here we shall give an argument that only uses the local form of *AC*. For each $a \in M$ the set of nodes of the apg Ma is in one-one correspondence with an ordinal, and the correspondence induces an apg structure on the ordinal. The resulting apg will be in the universe of well-founded sets and will be isomorphic to Ma. For each $a \in M$ let T_a be the class of apgs in the well-founded universe that are isomorphic to Ma' for some $a' \in M$ such that $a \equiv_M a'$. By the above each class T_a is non-empty and hence has elements of minimum possible rank in the well-founded universe. Let πa be the set of such elements of T_a. Note that if $a_1 \equiv_M a_2$ then $T_{a_1} = T_{a_2}$ so that $\pi a_1 = \pi a_2$. Conversely if $a_1, a_2 \in M$ such that $\pi a_1 = \pi a_2$ then there must be an apg in both T_{a_1} and T_{a_2}. Hence there must be $a_1', a_2' \in M$ such that $a_1 \equiv_M a_1'$, $a_2 \equiv_M a_2'$ and $Ma_1' \cong Ma_2'$. By exercise 2.8 $a_1' \equiv_M a_2'$ so that $a_1 \equiv_M a_2$. $\qquad\square$

2.18 Exercise: *(due to Dag Westerståhl) Show that*

$$AFA_1 \iff AFA_1^{ext},$$

where AFA_1^{ext} *is:*

Every extensional graph has at least one decoration.

2.19 Theorem: *The following are equivalent for each system M.*

(1) *M is strongly extensional.*

(2) *For each (small) system M_0 there is at most one system map $M_0 \longrightarrow M$.*

(3) *For each system M' every system map $M \longrightarrow M'$ is injective.*

Proof: We first show that (1) and (2) are equivalent. Assuming (1), let $\pi_1, \pi_2 : M_0 \longrightarrow M$ be system maps. By proposition 2.13(1) $(\pi_1 \times \pi_2)(=_{M_0})$ is a bisimulation R on M. If $m \in M_0$

then $(\pi_1 m)R(\pi_2 m)$ so that $\pi_1 m \equiv_M \pi_2 m$ and hence $\pi_1 m = \pi_2 m$, as M is strongly extensional. Thus $\pi_1 = \pi_2$ and we have proved (2). Now assume (2) and apply exercise 2.14, where R is the bisimulation $\equiv_M$, to construct the system M_0 and system maps $\pi_1, \pi_2 : M_0 \longrightarrow M$. By (2), $\pi_1 = \pi_2$, so that whenever $a \equiv_M b$ then $(a,b) \in M_0$ and $a = \pi_1(a,b) = \pi_2(a,b) = b$. Thus M is strongly extensional; i.e. (1).

We next show that (1) is equivalent to (3). Assume (1) and let $\pi : M \longrightarrow M'$ be a system map. By proposition 2.13(2) $(\pi \times \pi)^{-1}(=_{M'})$ is a bisimulation R on M. Hence if $\pi a = \pi b$, i.e. aRb, then $a \equiv_M b$ so that $a = b$, as M is strongly extensional. Thus π is injective and we have proved (3). Now assume (3) and by applying the previous lemma let $\pi : M \longrightarrow M'$ be a strongly extensional quotient of M. By (3) π must be injective and so an isomorphism $M \cong M'$. As M' is strongly extensional it follows that M is too.

Finally we show that the local version of (2), for small systems M_0 only, implies the unrestricted version. Let $\pi_1, \pi_2 : M_0 \longrightarrow M$ be system maps and let $a \in M_0$. Then by restricting π_1, π_2 to the small pointed system $M_0 a$ we may apply the restricted version of (2) to deduce that π_1 and π_2 are equal on $M_0 a$ so that $\pi_1 a = \pi_2 a$. As $a \in M$ was arbitrary it follows that $\pi_1 = \pi_2$. □

2.20 Proposition: *Let M be a system such that any two nodes of M lie in a common apg of the form Mc. Then M is strongly extensional iff Mc is strongly extensional for every node c of M.*

Proof: Observe that the identity map on Mc is an injective system map $Mc \longrightarrow M$. So two distinct system maps $M_0 \longrightarrow Mc$ would give rise to distinct system maps $M_0 \longrightarrow M$. Hence by (1) $\Rightarrow$ (2) of theorem 2.19 we get the implication from left to right. For the converse implication assume that Mc is strongly extensional for every node c of M. Let $\pi : M \longrightarrow M'$ be a system map and suppose that $a, b \in M$ such that $\pi a = \pi b$. Choose $c \in M$ such that $a, b \in Mc$. As Mc is strongly extensional (1) $\Rightarrow$ (3) of theorem 2.19 implies that π is injective on Mc, so that $a = b$. Thus π is injective on M. Hence by (3) $\Rightarrow$ (1) of theorem 2.19 M is strongly extensional. □

Applying this proposition to the system V we get the next characterisation of AFA_2.

2.21 Proposition:

$AFA_2 \iff$ *Every canonical picture is strongly extensional.*

Exact Pictures

We will call an apg an EXACT PICTURE if it has an injective decoration, i.e. distinct nodes are assigned distinct sets by the decoration. An alternative way to state this is to say that the apg is isomorphic to a canonical picture. Proposition 2.21 can be reformulated as stating that

$AFA_2 \iff$ Every exact picture is strongly extensional.

2.22 Proposition:

$AFA_1 \iff$ *Every strongly extensional apg is an exact picture.*

Proof: Assume AFA_1. Let G be a strongly extensional apg. By AFA_1 G has a decoration d. So $d : G \longrightarrow V$ is a system map. By $(1) \Rightarrow (3)$ of theorem 2.19 d is injective, so that G is an exact picture.

Conversely, let us assume the right hand side of the proposition and show that each graph G has a decoration. Given the graph G we may form an apg G' by adding a new node $*$ and new edges $(*, a)$ for each node a of G. Now let $\pi : G' \longrightarrow G''$ be a strongly extensional quotient of G'. Then $G''(\pi*)$ is strongly extensional and hence by our assumption it is an exact picture. So G'' has an injective decoration d''. Now d is a decoration of G where $da = d''(\pi a)$ for each node a of G.   □

Combining the characterisations of AFA_1 and AFA_2 that we have just obtained we get the main result.

2.23 Theorem: *AFA is equivalent to:*
    An apg is an exact picture iff it is strongly extensional.

The Normal Structure Axiom

Here we consider an axiom suggested by a completeness theorem in Kanger (1957) for a variant of the predicate calculus. This variant has atomic formulae of the form

$$(s_1, \ldots, s_n) \, \varepsilon \, t$$

where $n > 0$ and $s_1, \ldots, s_n, t$ are variables or individual constants. The natural semantics for the variant logic is to use structures $\mathcal{A} = (A, R, \ldots, c^{\mathcal{A}}, \ldots)$ where A is a non-empty set, $R \subseteq A^+ \times A$ and $c^{\mathcal{A}} \in A$ for each individual constant c. Here $A^+ = \bigcup_{n>0} A^n$. Let us call such a structure a KANGER structure. The standard completeness theorem will obviously carry over if this semantics is used. Kanger's idea is to modify the semantics by only using 'normal' Kanger structures in the definitions of logical validity and logical consequence. A NORMAL Kanger structure is a structure

$$\mathcal{A} = (A, R, \ldots, c^{\mathcal{A}}, \ldots)$$

where

$$R = \{(b, a) \in A^+ \times A \mid b \in a\}.$$

At first sight the restriction to normal structures may appear severe in view of the consistency of such sentences as

$$\exists x \, ((x, x) \, \varepsilon \, x).$$

In fact Kanger still succeeds in proving the variant logic complete relative to the normal structures. To do so he obviously has to invoke some principle that will imply the existence of enough non-well-founded sets to guarantee the existence of normal models of sentences such as the above one. Kanger formulates a set theoretical principle that is strong enough to imply that every countable Kanger structure is isomorphic to a normal such structure. In view of the Löwenheim-Skolem theorem this consequence implies that the standard completeness theorem will entail the completeness theorem relative to normal structures.

Here we will avoid cardinality considerations and formulate the following axiom:

The Normal Structure Axiom, *NSA*:
   Every Kanger structure is isomorphic to a normal one.

This axiom is certainly enough to give Kanger's completeness theorem. We have the following result.

2.24 Theorem: $AFA_1 \implies NSA$.

Proof: First note that it suffices to prove the result for Kanger structures (A, R); i.e. where there are no individual constants.

In order to apply AFA_1 define a graph G as follows. Let $\bar{A}$ be the smallest set such that $\{0\} \times A \subseteq \bar{A}$ and $\{1\} \times (\bar{A} \times \bar{A}) \subseteq \bar{A}$. Choose $\alpha \in On$ so that there is a bijection $f : A \to (\alpha - \{0\})$. Of course this requires AC. The nodes of G are the elements of the set $\bar{A} \cup (\{2\} \times \alpha)$. G has edges of the following forms:

(1) $(2, \beta) \to (2, \gamma)$ for $\gamma < \beta < \alpha$

(2) $(1, (x, y)) \to u$ for $x, y \in \bar{A}$ and $u \in \{x, y\}$

(3) $(0, a) \to \pi_n((0, a_1), \ldots, (0, a_n))$ for $((a_1, \ldots, a_n), a) \in R$

(4) $(0, a) \to (2, fa)$ for $a \in A$

To define $\pi_n : \bar{A}^n \to \bar{A}$ for $n = 1, 2, \ldots$ let

$$\pi(x, y) = (1, ((1, (x, x)), (1, (x, y)))).$$

Now let $\pi_1 x = x$ for $x \in A$ and let

$$\pi_{n+1}(x_1, \ldots, x_n, x_{n+1}) = \pi(\pi_n(x_1, \ldots, x_n), x_{n+1})$$

for $x_1, \ldots, x_n, x_{n+1} \in \bar{A}$.

By AFA_1 G has a decoration d. Note that the subgraph of G obtained by restricting to the nodes in $\{2\} \times \alpha$, is well-founded, having edges only of the form (1). It follows from the uniqueness part of Mostowski's collapsing lemma that

$$d(2, \beta) = \beta \quad \text{for all} \quad \beta < \alpha.$$

Also note that for all $x, y \in \bar{A}$.

$$d(1, (x, y)) = \{dx, dy\}$$

and hence

$$d(\pi(x, y)) = (dx, dy).$$

It follows that for all $x_1, \ldots, x_n \in \bar{A}$

$$d(\pi_n(x_1, \ldots, x_n)) = (dx_1, \ldots, dx_n).$$

Now let $\psi a = d(0, a)$ for $a \in A$. Then by considering the edges of G of the forms (3) and (4) we see that for all $a \in A$

(*) $\psi a = \{(\psi a_1, \ldots, \psi a_n) \mid ((a_1, \ldots, a_n), a) \in R\} \cup \{fa\}.$

We now make a sequence of observations:

(i) $dz \neq \emptyset$ for all $z \in G$ except $z = (2, 0)$.

(ii) $\emptyset \notin dz$ for all $z \in \bar{A}$.

(iii) $dz \notin On$ for all $z \in \bar{A}$.

(iv) fa is the unique ordinal in ψa for each $a \in A$.

(v) ψ is injective.

By $(*)$ and (v) it follows that $\psi : (A, R) \cong (B, S)$ where (B, S) is the normal Kanger structure with $B = \{\psi a \mid a \in A\}$. $\square$

The axiom NSA is a strengthening of a 'completeness' axiom considered in Gordeev (1982). For any set c let $V{\restriction}c$ be the graph having the elements of c as nodes and having edges $x \to y$ whenever $x \in y$ and $x, y \in c$. Call a graph of the form $V{\restriction}c$ a NORMAL graph. Then GORDEEV's axiom, GA, is:

Every graph is isomorphic to a normal one.

2.25 Exercise: *Show that*

$$NSA \implies GA.$$

3 | A Model of the Axiom

As in the previous chapters we shall work informally in the framework of the axiomatic set theory ZFC^-. The aim of this chapter is to form a class model of our set theory, including the new axiom AFA.

Complete Systems

Given a system M an M-DECORATION of a graph G is a system map $G \longrightarrow M$.

3.1 Example: *A V-decoration of G is simply a decoration of G.*

M is a COMPLETE system if every graph has a unique M-decoration. Note that by theorem 2.19 every complete system is strongly extensional. Also note that if M is strongly extensional and every strongly extensional graph has an M-decoration then M is complete. Finally, note that AFA holds if and only if the system V is complete.

We turn to the construction of a complete system. Every apg has the form Ga where G is a graph and a is a node of G. The class of apgs form a system V_0 with edges (Ga, Gb) wherever G is a graph and $a \longrightarrow b$ in G. Let $\pi_c : V_0 \longrightarrow V_c$ be a strongly extensional quotient of V_0.

3.2 Proposition:
For each system M there is a unique system map $M \longrightarrow V_c$

Proof: If $a \in M$ then $Ma \in V_0$. Moreover the map $M \longrightarrow V_0$ that assigns Ma to $a \in M$ is clearly a system map. Composing with the system map $\pi_c : V_0 \longrightarrow V_c$ we obtain a system map $M \longrightarrow V_c$. The uniqueness of this system map follows by theorem 2.19 from the strong extensionality of V_c. $\qquad \Box$

3.3 Corollary: *V_c is complete*

Proof: Apply the proposition to small systems M.   □

3.4 Theorem: *The following are equivalent for a system M.*

(1) *For each system M' there is a unique system map $M' \longrightarrow M$.*
(2) *M is complete.*
(3) *$M \cong V_c$.*

Proof: That (3) implies (1) is an immediate consequence of proposition 3.2. That (1) implies (2) is trivial. We now show that (2) implies (3). Let M be a complete system. Let $\pi : M \longrightarrow V_c$ be the unique system map which exists by proposition 3.2. The map π is injective as M is strongly extensional. If $a \in V_c$ then $V_c a$ is an apg with a unique M-decoration d say. Then $\pi \circ d : V_c a \longrightarrow V_c$ is a system map. As V_c is strongly extensional $\pi \circ d$ must be the identity map on $V_c a$. In particular $a = \pi(da)$. Thus π is surjective as well as injective. So $\pi : M \cong V_c$.   □

Full Systems

A system M is a FULL system if for every set $x \subseteq M$ there is a unique $a \in M$ such that $x = a_M$.

3.5 Example:

(1) *V is a full system. More generally whenever M is a class such that $M = \text{pow} M$ then M is a full system when $a \longrightarrow b$ in M iff $b \in a \in M$. For example the class V_{wf} of well-founded sets is such a full system. In fact V_{wf} is the smallest class M such that $M = \text{pow} M$. Note that V is the largest such class M and the foundation axiom can be expressed by the equation*

$$V = V_{wf}.$$

(2) *If $\pi : M \to M$ is any bijection on a full system M then we can obtain a new full system M_π having the same nodes as M but where*

$$a \to b \text{ in } M_\pi \iff \pi a \to b \text{ in } M.$$

3.6 Exercise: *Show that the following are equivalent for a full system M.*

- *For each full system M' there is a unique system map $M \longrightarrow M'$.*
- *M is well founded.*
- *$M \cong V_{wf}$.*

We will give two different proofs of the next result.

3.7 Proposition: *Each complete system is full.*

Proof 1: Let $x \subseteq M$ be a set, where M is a complete system. Form the graph G_0 that has the nodes and edges of M that lie on paths starting from a node in x. Let the graph G be obtained from G_0 by adding a new node $*$ and edges $(*, y)$ for each $y \in x$. As M is complete G has a unique M-decoration d. Restricting d to the nodes of G_0 we obtain an M-decoration of G_0. But the identity map is clearly the unique M-decoration of G_0. So $dx = x$ for $x \in G_0$. Hence if $a = d*$ then $a \in M$ such that

$$a_M = \{dy \mid * \to y \text{ in } G\}$$
$$= x.$$

Now suppose that $a' \in M$ such that $a'_M = x$. Then we get an M-decoration d' of G with $d'* = a'$ and $d'y = y$ for $y \in G_0$. As d is the unique M-decoration of G, $d = d'$ so that

$$a' = d'* = d* = a.$$

So we have shown that there is a unique $a \in M$ such that $a_M = x$.
□

Proof 2: Let M be a complete system. Observe that $pow M$ is a system, where if $x \in pow M$ then $x_{pow M} = \{y_M \mid y \in x\}$. As M is complete there is a unique system map $h : pow M \longrightarrow M$. So for all $x \in pow M$

$$(*) \quad (hx)_M = \{h(y_M) \mid y \in x\}.$$

Note that $(\ )_M : M \to pow M$ is also a system map, so that $h \circ (\ )_M : M \to M$ is a system map. But because M is complete the identity map on M is the unique system map $M \to M$. So

$$h(x_M) = x \quad for \ all \ x \in M.$$

Hence from $(*)$, for all $x \in pow M$

$$(hx)_M = \{hy_M \mid y \in x\}$$
$$= \{y \mid y \in x\}$$
$$= x.$$

Thus $h : pow M \to M$ and $(\ )_M : M \to pow M$ are inverses to each other, so that $(\ )_M$ is a bijection and hence M is full. □

The Interpretation of *AFA*

Any system M determines an interpretation of the language of set theory in which the variables range over the nodes of M and the predicate symbol '$\in$' is interpreted by the relation $\in_M$, where for $a, b \in M$

$$a \in_M b \iff a \in b_M.$$

When the system M is full this interpretation models all the axioms of ZFC^-. This fundamental result is due to Rieger (1957). A proof of Rieger's theorem may be found in appendix A.

3.8 Theorem:

> *Each complete system is a model of $ZFC^- + AFA$.*

Proof: Let M be a complete system. Then M is full and hence a model of ZFC^-, by Rieger's theorem. So it remains to prove that M is a model of AFA. If x is a subset of M then let x^M be the unique $a \in M$ such that $x = a_M$. For $a, b \in M$ let

$$(a, b)^{(M)} = \{\{a\}^M, \{a, b\}^M\}^M.$$

Then $(a, b)^{(M)}$ is the element of M that is the standard set theoretical representation in M of the ordered pair of a and b. Here we represent a graph as an ordered pair consisting of a set and a binary relation on it. So, for $c \in M$, $M \models$ "c is a graph" if and only if there are $a, b \in M$ such that $c = (a, b)^{(M)}$ and $M \models$ "b is a binary relation on a"; i.e. $b_M \subseteq \{(x, y)^{(M)} \mid x, y \in a_M\}$. Hence for such a $c \in M$ we may define a graph G having as nodes the elements of a_M and having as edges the pairs (x, y) such that $(x, y)^{(M)} \in b_M$. As M is complete G has a unique M-decoration. This is the unique map $d : a_M \to M$ such that for all $x \in a_M$

$$dx = \{dy \mid (x, y)^{(M)} \in b_M\}.$$

Now let $f = \{(x, dx)^{(M)} \mid x \in a_M\}^M$. Then $f \in M$ and it is a routine matter to check that

$$M \models \text{``}f \text{ is the unique decoration of the graph } c\text{''}.$$

Thus we have proved that in M every graph has a unique decoration; i.e. M is a model of AFA. □

3.9 Exercise: *Let M be a full system. Show that*

  (i) *M is a model of FA iff M is well-founded.*

  (ii) *M is a model of AFA_1 iff every graph has an M-decoration.*

(iii) *M is a model of AFA_2 iff M is strongly extensional.*

(iv) *M is a model of AFA iff M is complete.*

As an immediate consequence of part (iv) of this exercise and theorem 3.8 we get the following result.

3.10 Theorem: *$ZFC^- + AFA$ has a full model that is unique up to isomorphism.*

Part Two

Variants of the
Anti-Foundation Axiom

4 | Variants Using a Regular Bisimulation

In this chapter we shall consider two variants $FAFA$ and $SAFA$ of the axiom AFA. It turns out that all three axioms can be treated as different instances of a family of axioms $AFA^\sim$, one for each regular bisimulation $\sim$ having a definition that is absolute for full systems. What this means will be explained below. After presenting the general theory we shall consider each of the two variants in turn.

Recall that the system V_0 of apg's, defined before proposition 3.2 has an edge (Ga, Gb) whenever $a \longrightarrow b$ in the graph G. A bisimulation relation $\sim$ on V_0 is a REGULAR BISIMULATION relation if

(1) $\sim$ is an equivalence relation on V_0.

(2) $Ga \cong G'a' \implies Ga \sim G'a'$.

(3) $a_G = a'_G \implies Ga \sim Ga'$ for $a, a' \in G$.

4.1 Exercise: *Show that $\equiv_{V_0}$ is a regular bisimulation such that for any system M*

$$Ma \equiv_{V_0} Mb \iff a \equiv_M b.$$

We shall later give two other examples of regular bisimulations to get the two variants of AFA. Until then we assume given a fixed regular bisimulation $\sim$.

A system M is a $\sim$-EXTENSIONAL system if

$$Ma \sim Mb \implies a = b.$$

4.2 Exercise: *Show that if M is $\sim$-extensional then for any system M_0 there is at most one injective system map $M_0 \longrightarrow M$.*

41

Hint: Observe that if $\pi : M_1 \longrightarrow M_2$ is an injective system map then for $a \in M_1$

$$(\pi \upharpoonright M_1 a) \quad : \quad M_1 a \cong M_2(\pi a).$$

$\sim$-Complete Systems

A system M is a $\sim$-COMPLETE system if it is $\sim$-extensional and every $\sim$-extensional graph has an M-decoration. Note that by exercise 4.2 the M-decoration is necessarily unique.

4.3 Example: *If $\sim$ is $\equiv_{V_0}$ then*

- *M is $\sim$-extensional iff M is strongly extensional.*
- *M is $\sim$-complete iff M is complete.*

Our first aim is to construct a $\sim$-complete system. Let $V_0^{\sim}$ be the subsystem of V_0 consisting of the $\sim$-extensional apg's and all the edges of V_0 between such apg's. We let $V_c^{\sim}$ be a $\sim$-extensional system for which there is a surjective system map $\pi^{\sim} : V_0^{\sim} \longrightarrow V_c^{\sim}$ such that

$$Ga \sim G'a'; \quad \Longleftrightarrow \quad \pi(Ga) = \pi(G'a')$$

for all $\sim$-extensional apg's Ga and $G'a'$. We are guaranteed the existence of $V_c^{\sim}$ and $\pi^{\sim}$ by the following lemma.

4.4 Lemma: *For every system M there is a system M' and surjective system map $\pi : M \longrightarrow M'$ such that for $x, x' \in M$*

$$Mx \sim Mx' \quad \Longleftrightarrow \quad \pi x = \pi x'.$$

Moreover if Mx is $\sim$-extensional for all $x \in M$ then M' is $\sim$-extensional.

Proof: The first part is proved as in the proof of lemma 2.17. For the second part observe that for each $a \in M$ the restriction of π to Ma is a surjective system map

$$Ma \longrightarrow M'(\pi a).$$

Now suppose that $x, y \in Ma$ and $\pi x = \pi y$. Then $Mx \sim My$ so that by the $\sim$-extensionality of Ma it follows that $x = y$. Thus $\pi \upharpoonright Ma$ is an isomorphism $Ma \cong M'(\pi a)$ so that $Ma \sim M'(\pi a)$.

We can now show that M' is $\sim$-extensional. As $\pi : M \longrightarrow M'$ is surjective it suffices to show that

$$M'(\pi a) \sim M'(\pi b) \implies \pi a = \pi b.$$

But assuming that $M'(\pi a) \sim M'(\pi b)$ we get by the above that

$$Ma \sim M'(\pi a) \sim M'(\pi b) \sim Mb,$$

so that $Ma \sim Mb$ and hence $\pi a = \pi b$.   □

Note that in applying this lemma to $M = V_0^\sim$, if $x = Ga \in M$ then $Mx \cong Ga$ so that $Mx \sim Ga$ and Mx is $\sim$-extensional, as Ga is.

4.5 Proposition: *For each $\sim$-extensional system M there is a unique injective system map $M \longrightarrow V_c^\sim$.*

Proof: By exercise 4.2 the uniqueness of the injective system map follows from the fact that $V_c^\sim$ is $\sim$-extensional. So it only remains to show the existence of a system map $M \longrightarrow V_c^\sim$, where M is $\sim$-extensional. Clearly $\pi_M : M \longrightarrow V_0^\sim$ is a system map, where $\pi_M a = Ma$ for $a \in M$. Hence by composing with $\pi^\sim :$ $V_0^\sim \longrightarrow V_c^\sim$ we get a system map $\pi^\sim \circ \pi_M : M \longrightarrow V_c^\sim$. It only remains to show that this map is injective. So let $x, y \in M$ such that $\pi^\sim(Mx) = \pi^\sim(My)$. Then $Mx \sim My$ so that, as M is $\sim$-extensional, $x = y$.   □

4.6 Corollary: *$V_c^\sim$ is $\sim$-complete.*

Proof: We already know that $V_c^\sim$ is $\sim$-extensional. It only remains to apply the proposition to small systems M.   □

For systems M, M' let $M \preceq M'$ if there is an injective system map $M \longrightarrow M'$. Note that $\preceq$ is both reflexive and transitive. Our next aim is to prove the following result.

4.7 Theorem: *Let M be a $\sim$-extensional system. Then the following are equivalent:*

(1) M is $\sim$-complete.

(2) $M_0 \preceq M$ for every $\sim$-extensional system M_0.

(3) $M \preceq M' \implies M \cong M'$ for every $\sim$-extensional system M'.

(4) $M \cong V_c^\sim$.

Proof: For (1) implies (2) let M be $\sim$-complete and let M_0 be a $\sim$-extensional system. Then for $a \in M_0$ the apg $M_0 a$ must have an injective M-decoration d_a which, by exercise 4.2, is uniquely determined. Define $d : M_0 \longrightarrow M$ by

$$da = d_a a \quad for \quad a \in M_0.$$

Observe that if $a \in M$ then $d_x = d_a\, z(M_0 x)$ for $x \in a_M$, so that

$$(d_a a)_M = \{d_x x \mid x \in a_{M_0}\},$$

and hence $(da)_M = \{dx \mid x \in a_{M_0}\}$. Thus d is a system map. To see that d is injective use the hint to exercise 4.2 to get that

$$d_x : M_0 x \cong M(dx) \quad and \quad d_y : M_0 y \cong M(dy),$$

so that if $dx = dy$ then $M_0 x \cong M_0 y$. It follows that if $dx = dy$ then $M_0 x \sim M_0 y$ and hence $x = y$, as M_0 is $\sim$-extensional.

For (2) implies (3) let $M \preceq M'$, where M' is $\sim$-extensional. By (2) $M' \preceq M$. So there are injective system maps $M \longrightarrow M'$ and $M' \longrightarrow M$. By exercise 4.2 their compositions must be the identity maps on M and M' so that $M \cong M'$.

For (3) implies (4) use proposition 4.5 to get that $M \preceq V_c^\sim$. As $V_c^\sim$ is $\sim$-extensional we may apply (3) to get (4).

For (4) implies (1) apply Corollary 4.6.   □

The next result generalises proposition 3.7.

4.8 Lemma: *Every $\sim$-complete system is full.*

Proof: Let $x \subseteq M$ be a set, where M is a $\sim$-complete system. As in the proof of proposition 3.7, we may form the graph G_0 consisting of the nodes and edges of M that lie on paths starting from a node in x. Again we may let G be obtained from G_0 by adding a new node $*$ and edges $(*, y)$ for $y \in x$. If G is $\sim$-extensional then by taking the unique M-decoration of G we can argue as before. But if G is not $\sim$-extensional then, as G_0 is, it must be the case that $G* \sim Ga$ for some $a \in G_0$. As $\sim$ is a bimulation it follows that

$$\forall y \in {}_{*G} \exists a' \in a_G \ Gy \sim Ga' \quad \& \quad \forall a' \in a_G \exists y \in {}_{*G} \ Gy \sim Ga'.$$

But $*_G = x$ and $a \in M$ with $a_G = a_M$. Also $Gy = My$ for $y \in x$ and $Ga' = Ma'$ for $a' \in a_G$. Thus

$$\forall y \in x \exists a' \in a_M \ My \sim Ma' \quad \& \quad \forall a' \in a_M \exists y \in x \ My \sim Ma'.$$

Hence, as M is $\sim$-extensional

$$x = a_M.$$

The uniqueness of a is a consequence of the fact that M is $\sim$-extensional and hence extensional.   □

The Axioms $AFA^\sim$

So far we have not given our generalisation of AFA. To do so we must assume given a definition in the language of set theory of the regular bisimulation $\sim$. So we assume given a formula $\phi(x,y)$, without any parameters and having at most the variables x, y free, that defines $\sim$ in V. This means that for all apg's c and d

$$c \sim d \iff V \models \phi(c,d).$$

We shall assume that $\phi(x,y)$ is fixed and refer to it as the definition of $\sim$. Using the definition of $\sim$ we may form a sentence that expresses that V is $\sim$-complete. Let us call this sentence $AFA^\sim$. It is this sentence that is our generalisation of AFA. Note that in case $\sim$ is $\equiv_{V_0}$ then

$$AFA^\sim \iff AFA.$$

As with AFA we may split $AFA^\sim$ into its two parts

- $AFA_1^\sim$: Every $\sim$-extensional graph has an injective decoration.
- $AFA_2^\sim$: V is $\sim$-extensional.

The following is straightforward to prove.

4.9 Proposition:
(1) $AFA_1^\sim$ iff every $\sim$-extensional apg is an exact picture.
(2) $AFA_2^\sim$ iff every exact picture is $\sim$-extensional.

4.10 Corollary: $AFA^\sim$ is equivalent to: an apg is an exact picture iff it is $\sim$-extensional.

A result that we have not yet generalised is theorem 3.8; i.e. that each complete system M is a full model of $ZFC^- + AFA$. To do so we need to assume that the definition of $\sim$ is absolute for full systems. To spell out what this means let M be a full system and let $\sim_M$ be the relation on M that the definition $\phi(x,y)$ of $\sim$ defines in M; i.e. for $c, d \in M$

$$c \sim_M d \iff M \models \phi(c,d).$$

For each $c \in M$ such that $M \models$ "c is an apg" there is a natural way to obtain an apg from it (see below). Let us call the result $ext_M(c)$. The formula $\phi(x,y)$ is an ABSOLUTE formula for M if for all $c, d \in M$ such that $M \models$ "c, d are apg's"

$$c \sim_M d \iff ext_M(c) \sim ext_M(d).$$

We turn to the definition of $ext_M(c)$. A pointed graph will here be represented as a triple $((a,b),u)$ where a is a set, b is a binary relation on a and u is an element of a. So if $c \in M$ then $M \models$ "c is a pointed graph" if and only if $c = ((a,b)^{(M)}, u)^{(M)}$ for some (uniquely determined) $a, b, u \in M$ such that

$$b_M \subseteq \{(x,y)^{(M)} \mid x, y \in a_M\}$$

and $u \in a_M$. With such a c in M we may associate the pointed graph

$$((a_M, \{(x,y) \mid (x,y)^{(M)} \in b_M\}), u).$$

Call this $ext_M(c)$.

We can now state the generalisation of theorem 3.8.

4.11 Theorem: *Let $\sim$ be a regular bisimulation whose definition is absolute for full systems. Then each $\sim$-complete system M is a full model of $ZFC^- + AFA^\sim$.*

Part (iv) of exercise 3.9 also generalises, so that we get the final general result.

4.12 Theorem: *Let $\sim$ be a regular bisimulation whose definition is absolute for full systems. Then $ZFC^- + AFA^\sim$ has a full model that is unique up to isomorphism.*

Finsler's Anti-Foundation Axiom

In this section we apply the general theory of the previous section to an axiom inspired by Finsler (1926). In that paper Finsler presents three axioms for a universe consisting of a collection of objects, to be called sets, and a binary relation $\in$ between them. His axioms are as follows:

I. $\in$ is decidable.

II. Isomorphic sets are equal.

III. The universe has no proper extension that satisfies I. and II.

If we take Finsler's universe to be a system in our sense then we can ignore axiom I and turn to his axiom II. One might expect that the correct way to express Finsler's notion of isomorphism in a system M is to take $a, b \in M$ to be isomorphic if the apg's Ma and Mb that they determine are isomorphic apg's. According to this view M is a model of II. iff it is $\cong$-extensional; i.e.

$$Ma \cong Mb \implies a = b.$$

But on examining Finsler's paper this is clearly seen to be incorrect. In fact Finsler understands his axiom II to be a strengthening of the extensionality axiom. But $\cong$-extensional systems need not be extensional. For example consider the two element graph G:

It has nodes a and b and edges (a, b) and (b, b). Clearly $Ga \ncong Gb$ but $a_G = \{b\} = b_G$. So G is $\cong$-extensional but not extensional.

A correct formulation of Finsler's notion of isomorphism will be given using the following construction. If $a \in M$, where M is a system, let $(Ma)^*$ be the apg consisting of the nodes and edges of Ma that are on paths starting from some child of a, together with a new node $*$ and a new edge $(*, x)$ for each child x of a. We take $*$ to be the point of $(Ma)^*$. Note that if a does not lie on any path starting from a child of a then $(Ma)^*$ will be isomorphic to Ma via an isomorphism that is the identity except that $*$ is mapped to a. If a does lie on such a path then $(Ma)^*$ consists of the nodes and edges of Ma together with the new nodes and edges.

We define $a, b \in M$ to be isomorphic in Finsler's sense if $(Ma)^* \cong (Mb)^*$. Note that if $a_M = b_M$ then $(Ma)^* = (Mb)^*$ and hence $(Ma)^* \cong (Mb)^*$.

Let $\cong^*$ be the relation on V_0 defined by:

$$Ga \cong^* G'a' \quad \Longleftrightarrow \quad (Ga)^* \cong (G'a')^*.$$

We call a system M a FINSLER-EXTENSIONAL system if it is $\cong^*$-extensional; i.e.

$$Ma \cong^* Mb \quad \Longrightarrow \quad a = b.$$

It is the Finsler-extensional systems that we take to be the models of axiom II.

4.13 Exercise: *Show that*

(i) $\cong^*$ *is a regular bisimulation.*

(ii) *A system M is Finsler-extensional iff it is both extensional and $\cong$-extensional.*

4.14 Exercise: *Let $\sim$ be the relation on V_0: $Ga \sim G'a'$ iff there is a bijection $\psi : a_G \cong a'_{G'}$ such that $Gx \cong G'(\psi x)$ for $x \in a_G$. Show that*

(i) $Ga \cong^* G'a' \quad \Longrightarrow \quad Ga \sim G'a'.$

(ii) $\sim$ *is a regular bisimulation.*

(iii) *M is $\sim$-extensional iff M is Finsler-extensional.*

Let us now consider Finsler's axiom III. I take a Finsler-extensional system to be a model of axiom III if any injective system map $M \longrightarrow M'$ is an isomorphism if M' is a Finsler-extensional system. By theorem 4.7 a system M is a model of Finsler's axioms iff M is Finsler-complete (i.e. M is $\cong^*$-complete).

4.15 Exercise: *Show that $\cong^*$ has a definition that is absolute for full systems.*

By this result we may form the axiom $AFA^{\cong^*}$, which we will call FINSLER'S ANTI-FOUNDATION AXIOM, or $FAFA$ for short. The previous work applies to give us the following two results.

4.16 Theorem: *FAFA is equivalent to:*
An apg is an exact picture iff it is Finsler-extensional.

4.17 Theorem: *$ZFC^- + FAFA$ has a full model that is unique up to isomorphism.*

Scott's Anti-Foundation Axiom

In Scott (1960) a model of ZFC^- with non-well-founded sets is constructed out of irredundant trees. Scott defines a tree to be a REDUNDANT tree if it has a proper automorphism; i.e. an automorphism that moves some node. The tree is an IRREDUNDANT tree otherwise. Scott (1960) gives another characterisation of this notion. We leave this as an exercise.

4.18 Exercise: *Show that a tree Tr is redundant iff there is a node c of Tr and distinct $a, b \in c_T$ such that $Ta \cong Tb$.*

Scott's idea is to use irredundant trees to represent the structure of sets. Recall that the canonical tree picture of a set c is obtained by unfolding the canonical picture Vc of c. Scott's model construction may be described as follows. Let V_0^t be the subsystem of V_0 consisting of the irredundant trees with all the edges of V_0 between such nodes. A system V_c^t and a surjective system map $\pi : V_0^t \longrightarrow V_c^t$ are constructed so that for trees $Tr, T'r'$

$$\pi(Tr) = \pi(T'r') \quad \Longleftrightarrow \quad Tr \cong T'r'.$$

V_c^t can be shown to be full and hence a model of ZFC^-. Moreover it is also a model of

- A tree is isomorphic to a canonical tree picture iff it is irredundant.

We shall call this SCOTT'S ANTI-FOUNDATION AXIOM, or $SAFA$ for short. It is essentially the axiom formulated in Scott (1960).

In the rest of this section we will show that the axiom $SAFA$ and its full model V_c^t are really special cases of the axiom $AFA^\sim$ and its model $V_c^\sim$ for a suitable choice of the regular bisimulation $\sim$. For any apg Ga let $(Ga)^t$ denote its unfolding. So the nodes of $(Ga)^t$ are the finite paths of Ga that start from a. Let $\cong^t$ be the relation on V_0 given by

$$Ga \cong^t G'a' \quad \Longleftrightarrow \quad (Ga)^t \cong (G'a')^t.$$

4.19 Exercise: *Show that $\cong^t$ is a regular bisimulation which has a definition that is absolute for full models.*

By this result we may obtain the axiom $AFA^{\cong^t}$ and its model $V_c^{\cong^t}$. The next three results will be needed to show that $AFA^{\cong^t}$ is equivalent to $SAFA$.

4.20 Lemma:
The unfolding of a $\cong^t$-extensional apg is an irredundant tree.

Proof: Let Gn be a $\cong^t$-extensional apg. Let $a, b \in c_G$, where $c \in (Gn)^t$, such that $(Gn)^t a \cong (Gn)^t b$. Then $(Ga)^t = (Gn)^t a \cong (Gn)^t b = (Gb)^t$ so that $Ga \cong^t Gb$ and hence $a = b$ as G is $\cong^t$-extensional. Thus $(Gn)^t$ is irredundant.   □

4.21 Lemma: *If Tr is an irredundant tree then there is a $\cong^t$-extensional apg Gn and a surjective system map $\pi : Tr \longrightarrow Gn$ such that $Tr \cong (Gn)^t$ and for $a, b \in Tr$*

$$\pi a = \pi b \quad \Longleftrightarrow \quad Ta \cong Tb.$$

Proof: Let Tr be an irredundant tree. Let $\sim$ be the equivalence relation on the nodes of Tr defined by

$$a \sim b \quad \Longleftrightarrow \quad Ta \cong Tb,$$

for $a, b \in Tr$. As $\sim$ is a bisimulation equivalence we can form a quotient $\pi : Tr \to Gn$ of Tr with respect to $\sim$ by letting

$$\pi a = \{ b \in Tr \mid a \sim b \}$$

for $a \in Tr$, and letting $G = \{ \pi a \mid a \in Tr \}$ and $n = \pi r$. It only remains to show that $Tr \cong (Gn)^t$. So define $\psi : Tr \to (Gn)^t$ by:

$$\psi a = (\pi r, \dots, \pi a)$$

for $a \in Tr$, where $r \to \cdots \to a$ is the unique path in Tr between the root r and the node a. That ψ is a subjective system map should be clear. To see that ψ is injective let $a, b \in Tr$ such that $\psi a = \psi b = (n, \dots, c)$. Then there are paths $r \to \cdots \to a$ and $r \to \cdots \to b$ in Tr such that $\pi r = n, \cdots, \pi a = \pi b = c$. Suppose that $a \neq b$. Then there is a first node c' in the path $n \to \cdots \to c$ whose corresponding nodes a' and b' in the paths $r \to \cdots \to a$ and $r \to \cdots \to b$ are distinct, even though $\pi a' = \pi b' = c'$. Then $Ta' \cong Tb'$ and a' and b' are children of the common node that precedes them in the paths $r \to \cdots \to a' \to \cdots \to a$ and $r \to \cdots \to b' \to \cdots \to b$. So, as Tr is irredundant $a' = b'$, contradicting the choice of a' and b'. Hence we must have $a = b$. Thus ψ is injective and hence $\psi : Tr \cong (Gn)^t$.   □

4.22 Lemma: *If Ga and $G'a'$ are $\cong^t$-extensional apg's then*

$$Ga \cong^t G'a' \implies Ga \cong G'a'.$$

Proof: Let Ga and $G'a'$ be $\cong^t$-extensional apg's such that $Ga \cong^t G'a'$. Then by lemma 4.20 $(Ga)^t$ and $(G'a')^t$ are isomorphic irredundant trees. Let $\psi : (Ga)^t \cong (G'a')^t$. Define $\pi : Ga \to G'a'$ as follows: If $b \in Ga$ let σ be a path from a to b in Ga. Then $\sigma \in (Ga)^t$ so that $\psi\sigma \in (G'a')^t$. Let πb be the last node c in $G'a'$ of the path $\psi\sigma$. To see that πb is well-defined let σ' also be a path from a to b in Ga and let c' be the last node in $G'a'$ of $\psi\sigma'$. Observe that the subtrees of $(Ga)^t$ determined by the two paths σ and σ' will both be isomorphic to the tree $(Gb)^t$ and hence to each other. It follows that the corresponding subtrees of $G'a'$ determined by $\psi\sigma$ and $\psi\sigma'$ will also be isomorphic. But these are isomorphic to $(G'c)^t$ and $(G'c')^t$ so that $(G'c)^t \cong (G'c')^t$ and hence $G'c \cong^t G'c'$. As G' is $\cong^t$-extensional $c = c'$. Thus π is well-defined and a similar argument shows that π is injective. That π is also surjective and is a system map should be routine to check. $\qquad\square$

The axiom $SAFA$ may be split into the two parts:

- $SAFA_1$: Every irredundant tree is isomorphic to a canonical tree picture.
- $SAFA_2$: Every canonical tree picture is irredundant.

4.23 Theorem:

(1) $SAFA_2 \iff AFA_2^{\cong^t}$.

(2) $SAFA \implies AFA_1^{\cong^t} \implies SAFA_1$.

(3) $SAFA \iff AFA^{\cong^t}$.

Proof: First note that (3) is an immediate consequence of (1) and (2). We now prove the four implications that make up (1) and (2).

- $SAFA_2 \implies AFA_2^{\cong^t}$.
  Let a, b be sets and let $c = \{a, b\}$. Then by $SAFA_2$ the tree $(Vc)^t$ is irredundant. As a, b determine children of the common node c of $(Vc)^t$

$$(Va)^t \cong (Vb)^t \implies a = b.$$

Thus V is $\cong^t$-extensional and so $AFA_2^{\cong^t}$ is proved.

- $AFA^{\cong^t} \implies SAFA_2$.
  By $AFA_2^{\cong^t}$ the apg Va is $\cong^t$-extensional so that by lemma 4.20 the tree $(Va)^t$ is irredundant.

- $SAFA \implies AFA_1^{\cong^t}$.
  Let Ga be a $\cong^t$-extensional apg. Then by lemma 6.1 the tree $(Ga)^t$ is irredundant. Hence by $SAFA_1$ there is a set c such that $(Ga)^t \cong (Vc)^t$. By (1) it follows from $SAFA_2$ that Vc is $\cong^t$-extensional. Hence by lemma 4.22 $Ga \cong Vc$. Thus Ga is an exact picture of c.

- $AFA_1^{\cong^t} \implies SAFA_1$.
  Let Tr be an irredundant tree and let $\pi : Tr \longrightarrow Gn$ be as in lemma 4.21 so that Gn is $\cong^t$-extensional and $T \cong (Gn)^t$. By $AFA_1^{\cong^t}$ there is a set c such that $Gn \cong Vc$ so that $Tr \cong (Ga)^t \cong (Vc)^t$. Thus Tr is isomorphic to a canonical tree picture. □

4.24 Theorem: *SAFA is equivalent to:*

An apg is an exact picture iff it is Scott extensional.

4.25 Theorem: *$ZFC^- + SAFA$ has a full model that is unique up to isomorphism.*

The Relationship Between the $AFA^\sim$

We have considered three examples of regular bisimulations $\sim$ that have a definition that is absolute for full systems. In each case we get an axiom $AFA^\sim$ which has a unique full model up to isomorphism. The three relations are $\equiv_{V_0}$, $\cong^*$, $\cong^t$ and these determine the axioms AFA, $FAFA$ and $SAFA$ respectively. What is the relationship between these axioms? The following proposition summarises what we know so far. The axioms AFA and $FAFA$ are at opposite extremes of the family of axioms $AFA^\sim$. While AFA expresses that only the strongly extensional apg's are exact pictures $FAFA$ expresses that any Finsler-extensional apg is an exact picture. The axiom $SAFA$ fits somewhere in between and it is the aim of this section to show that it fits strictly in between the two extremes.

4.26 Proposition: *Let $\sim$ be a regular bisimulation having a definition that is absolute for full models. Then*

(1) *Every strongly extensional system is ~-extensional.*

(2) *Every ~-extensional system is Finsler-extensional.*

(3) $AFA_2 \implies AFA_2^\sim \implies FAFA_2$.

(4) $FAFA_1 \implies AFA_1^\sim \implies AFA_1$.

(5) *If (a): There is a ~-extensional system that is not strongly extensional then*
$$\neg(AFA_1^\sim \& AFA_2).$$

(6) *If (b): There is a Finsler-extensional system that is not ~-extensional then*

$$\neg(FAFA_1 \& AFA_2^\sim).$$

(7) *If both (a) and (b) then the axioms FAFA, AFA and AFA^~ are pairwise incompatible.*

4.27 Theorem:

(1) There is a $\cong^t$-extensional graph that is not strongly extensional.

(2) There is a Finsler-extensional graph that is not $\cong^t$-extensional.

Proof:

(1) Consider the graph G:

with the distinct nodes a, b. This is $\cong^t$-extensional because $(Ga)^t \not\cong (Gb)^t$. In fact $(Ga)^t$ is

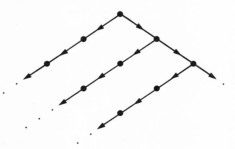

while $(Gb)^t$ is simply

But G is clearly not strongly extensional. Note that assuming AFA, Ga is a non-exact picture of Ω and Gb is an exact picture of Ω. On the other hand if we assume $SAFA$ then Gb is still an exact picture of Ω but Ga is an exact picture of a set $T \neq \Omega$ such that $T = \{\Omega, T\}$.

(2) This time let G be the graph:

with the distinct nodes a, b, c. Note that the unfolding $(Ga)^t$ of the apg Ga has the diagram:

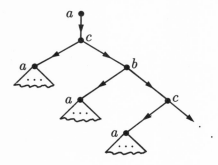

where the nodes of the tree have been labelled with the names of the corresponding nodes of G. It is clear from this diagram that the subtrees $(Gb)^t$ and $(Gc)^t$ are isomorphic. This shows that G is not $\cong^t$-extensional. But G is clearly extensional. Also it is rigid in the sense that it only has the identity automorphism. As every node is accessible from every other node it follows that G is Finsler-extensional. Note that assuming AFA the unique decoration of G will assign the set Ω to every node. But when $SAFA$ is assumed there is a decoration of G in which the nodes b and c get assigned a set X and the node a gets assigned a distinct set Y such that $Y = \{X\}$ and $X = \{X, Y\}$. Finally if $FAFA$ is assumed there is a unique injective decoration that assigns pairwise distinct sets A, B, C to the nodes a, b, c respectively so that $A = \{C\}, B = \{A, C\}$ and $C = \{B, C\}$.   □

When this theorem was presented in the course I only had an infinite example for part 2. The first finite example was found by Randoll Dougherty after I raised the problem in a talk at Berkeley. His graph had 9 nodes and 26 edges and after a series of improvements the above simple example with only 3 nodes and 5 edges was found by Larry Moss. Another example with the same number of nodes and edges was found independently by Scott Johnson. It is the following graph:

4.28 Corollary:
AFA, $FAFA$ and $SAFA$ are pairwise *incompatible axioms*.

5 | Another Variant

Boffa's Weak Axiom

Recall that Vza is the subgraph of the system V having all the nodes in the set a and all the edges of V between those nodes. Note that Vza need not be extensional but certainly is when a is a transitive set; i.e. when $x \in a$ always implies that $x \subseteq a$. Let BA_1 be the following axiom:

- Every extensional graph is isomorphic to Vza for some transitive set a.

Notice the similarity between this axiom and Gordeev's axiom GA mentioned at the end of chapter 2. The statement of GA involves a weakening of both the hypothesis and the conclusion of the above statement of BA_1. In fact BA_1 is a good deal stronger than GA. We saw in chapter 2 that $AFA_1 \implies GA$. Here we will see in 5.3 that BA_1 is strictly stronger than AFA_1.

Call a set x a REFLEXIVE set if $x = \{x\}$. As any two reflexive sets are isomorphic it follows from $FAFA_2$, and hence from any $AFA^\sim$, that there is at most one reflexive set. This is in sharp contrast to the situation when BA_1 is assumed. For example, by considering the two-element extensional graph

we obtain a two-element set of reflexive sets. A set of reflexive sets of any cardinality can be obtained equally easily, so that we have:

5.1 Proposition:
   Assuming BA_1 the reflexive sets form a proper class.

BA_1 may be viewed as giving the most generous possible answer to the question

Which apgs are exact pictures?

5.2 Proposition: BA_1 *is equivalent to:*

an apg is an exact picture iff it is extensional.

Proof: Note that every exact picture must be extensional by the axiom of extensionality. Let Ga be an extensional apg. Then by BA_1 there must be a transitive set c and $b \in c$ such that $Ga \cong (Vzc)b \cong Vb$. Hence Ga is an exact picture. Conversely suppose that every extensional apg is an exact picture and let G be an extensional graph. There are two cases. First suppose that $a_G = G$ for some $a \in G$. Then Ga is an extensional apg containing all the nodes of G. As Ga is extensional it is an exact picture so that $Ga \cong Vc$ for some set c. As $a_G = G$ it follows that c must be a transitive set and that $c \in c$ so that $G \cong Vzc$. Second suppose that $a_G \neq G$ for all $a \in G$. Then we can form an extensional apg $G'*$ by adding a new node $*$ to G and new edges $(*, a)$ for each $a \in G$. As $G'*$ is extensional it is an exact picture so that $G'* \cong Vc$ for some set c. As $a \in *_{G'}$ implies that $a_G \subseteq *_{G'}$ it follows that c must be a transitive set and $G \cong Vzc$.

$\square$

5.3 Corollary:
$BA_1 \implies AFA_1^\sim$ & $\neg AFA_2^\sim$ for any regular bisimulation $\sim$.

5.4 Exercise: *Show that every graph is isomorphic to a subgraph of an extensional graph.*

Recall from chapter 4 that $M_0 \preceq M$ if there is an injective system map $M_0 \longrightarrow M$. Note that $G \preceq V$ iff $G \cong (Vzc)$ for some transitive set c. Call an extensional system M a LOCALLY UNIVERSAL system if $G \preceq M$ for every extensional graph G. Observe that BA_1 is equivalent to:

- V is locally universal.

5.5 Exercise: *Show that a full system is a model of BA_1 iff it is locally universal.*

Boffa's Axiom and Superuniversal Systems

Here we will consider an axiom for non-well-founded sets due to M. Boffa. Assuming that $V \cong On$ we will show that this axiom has a unique full model up to isomorphism. In this respect it is like the axioms $AFA^\sim$, but it turns out not to be one of these.

The following notion will be useful when we come to formulate Boffa's strengthening of BA_1. The system M is a TRANSITIVE SUBSYSTEM of the system M', abreviated $M \trianglelefteq M'$, if $M \subseteq M'$ and

$$x_{M'} = x_M \quad \text{for all } x \in M.$$

5.6 Exercise: *Show that*

(i) *$M \trianglelefteq M'$ iff $M \subseteq M'$ and the inclusion map $M \hookrightarrow M'$ is a system map.*

(ii) *$G \trianglelefteq V$ iff $G = (V {\restriction} c)$ for some transitive set c.*

(iii) *If $M_i \trianglelefteq M$ for $i \in M$ then $\bigcup_{i \in I} M_i \trianglelefteq M$, where $\bigcup_{i \in I} M_i$ is the subsystem of M that has the nodes and edges of M that are in some M_i.*

(iv) *Every injective system map $M \longrightarrow M'$ has a unique factorisation*

$$M \longleftrightarrow M_0 \hookrightarrow M'$$

where $M_0 \trianglelefteq M'$. Here we use $M \longleftrightarrow M'$ to denote an isomorphism between M and M'.

(v) *Every injective system map $G \longrightarrow G'$ has a factorisation*

$$G \hookrightarrow G_0 \longleftrightarrow G'.$$

Let us now formulate what we shall call BOFFA'S ANTI-FOUN-DATION AXIOM, or *BAFA* for short:

- Every exact decoration of a transitive subgraph of an extensional graph can be extended to an exact decoration of the whole graph.

We may use arrow diagrams to express this axiom as follows:

- In the category of extensional systems and injective system maps the diagram

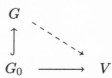

can always be completed. This means that given extensional graphs G_0 and G with $G_0 \trianglelefteq G$ and an injective system map $G_0 \longrightarrow V$ there is an injective system map $G \longrightarrow V$ that makes the diagram above commute.

5.7 Proposition: *For any extensional system M the following are equivalent:*

(1) *Any diagram*

 can be completed.

(2) *Any diagram*

 can be completed.

(3) *Any diagram*

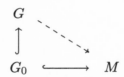

 can be completed.

In these diagrams G_0 and G are extensional graphs and all arrows denote injective system maps.

Proof: The implications (1) implies (2) and (2) implies (3) are trivial. For (3) implies (1), given maps $G_0 \longrightarrow G$ and $G_0 \longrightarrow M$ these maps may be factorised using part (iv) of exercise 5.6 to get

$$G_0 \longleftrightarrow G_0'' \hookrightarrow G \text{ and } G_0 \longleftrightarrow G_0' \hookrightarrow M.$$

Composing the two isomorphisms we get an isomorphism $G_0'' \longleftrightarrow G_0'$ and hence a map $G_0' \longrightarrow G$ which may be factorised, using part (v) of the same exercise, to give

$$G_0' \hookrightarrow G' \longleftrightarrow G.$$

By (3) the diagram

can be completed. Hence we get the following commutative diagram

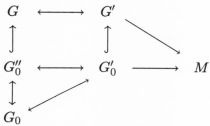

From this diagram we get a map $G \longrightarrow M$ which completes the diagram

$$
\begin{array}{ccc}
G & \dashrightarrow & \\
\uparrow & & \searrow \\
G_0 & \longrightarrow & M
\end{array}
$$

so that (1) is proved.   □

If the conditions in this proposition hold then we say that M is a SUPERUNIVERSAL system. Note that

$$BAFA \iff V \text{ is superuniversal}.$$

5.8 Exercise: *Show that a full system M is a model of BAFA iff it is superuniversal.*

5.9 Theorem: *Every superuniversal system is full.*

Proof: Let M be a superuniversal system and let $x \subseteq M$ be a set. We must find $a \in M$ such that $x = a_M$. The uniqueness of a follows from the fact that M is extensional. Let G_0 be the graph consisting of the nodes and edges of M that lie on paths starting from an element of x. Let G consist of the nodes and edges of G_0 together with a new node $*$ and new edges $(*, y)$ for $y \in x$. There are two cases to consider.

In the first case suppose that G is extensional. Then by the superuniversality of M the diagram

$$
\begin{array}{ccc}
G & \dashrightarrow & \\
\uparrow & & \searrow \\
G_0 & \hookrightarrow & M
\end{array}
$$

can be completed with an injective system map $d : G \longrightarrow M$. As d is the identity on G_0 and $*_G = x$, if $a = d*$ then

$$a_M = \{dy \mid y \in *_G\} = x.$$

In the second case suppose that G is not extensional. As $G_0 \trianglelefteq M$ and M is extensional it follows that G_0 is extensional. So there must be $a \in G_0$ such that $a_G = *_G$. But $*_G = x$ and $G_0 \trianglelefteq G$ and $G_0 \trianglelefteq M$ so that

$$a_M = a_{G_0} = a_G = *_G = x.$$

$\square$

5.10 Exercise: *(See Boffa 1972a) Show that BAFA implies σ, where σ expresses that for every set x there is a set y distinct from x such that $y = \{x, y\}$. Show that BA_1 does not imply σ by finding a globally universal* full system that is not a model of σ.*

A Backwards and Forwards Argument

In the rest of this chapter we show that there is a unique superuniversal system up to isomorphism. The next lemma will give us the backwards and forwards step in a transfinite backwards and forwards construction of an isomorphism between two superuniversal systems.

5.11 Lemma: *Assume given a diagram*

$$
\begin{array}{ccc}
G & \hookrightarrow & M \\
\updownarrow & & \\
G' & \hookrightarrow & M'
\end{array}
$$

---

* The notion of a globally universal system is defined just before 5.13 below.

If M' is superuniversal and $m \in M$ then there are graphs $F \trianglelefteq M$ and $F' \trianglelefteq M'$ and an isomorphism $F \longleftrightarrow F'$ such that $m \in F$ and the diagram

$$
\begin{array}{ccccc}
G & \lhook\joinrel\longrightarrow & F & \lhook\joinrel\longrightarrow & M \\
\big\updownarrow & & \big\uparrow & & \\
G' & \lhook\joinrel\longrightarrow & F' & \lhook\joinrel\longrightarrow & M'
\end{array}
$$

commutes. Moreover if M is also superuniversal and $m' \in M'$ then F and F' can be found as above so that also $m' \in F'$.

Proof: As $G \trianglelefteq M$ and $(Mm) \trianglelefteq M$ it follows that $G \cup (Mm) \trianglelefteq M$. Let $F = G \cup (Mm)$. Then $m \in F$. As M' is superuniversal the diagram

$$
\begin{array}{ccc}
G & \lhook\joinrel\longrightarrow & F \\
\big\updownarrow & & \Big\downarrow \\
G' & \lhook\joinrel\longrightarrow & M'
\end{array}
$$

can be completed with an injective system map $F \longrightarrow M'$, which can be factorised to give $F \longleftrightarrow F' \hookrightarrow M'$ and hence the diagram

$$
\begin{array}{ccccc}
G & \lhook\joinrel\longrightarrow & F & \lhook\joinrel\longrightarrow & M \\
\big\updownarrow & & \big\updownarrow & & \\
G' & \lhook\joinrel\longrightarrow & F' & \lhook\joinrel\longrightarrow & M'
\end{array}
$$

If M is also superuniversal and $m \in M'$ then we can repeat this construction starting from the diagram

$$
\begin{array}{ccc}
F & \lhook\joinrel\longrightarrow & M \\
\big\updownarrow & & \\
F' & \lhook\joinrel\longrightarrow & M'
\end{array}
$$

except that the roles of M and M' are interchanged. This time we get the commuting diagram

$$
\begin{array}{ccccc}
F & \lhook\joinrel\longrightarrow & H & \lhook\joinrel\longrightarrow & M \\
\big\updownarrow & & \big\uparrow & & \\
F' & \lhook\joinrel\longrightarrow & H' & \lhook\joinrel\longrightarrow & M'
\end{array}
$$

and hence the commuting diagram

$$
\begin{array}{ccccccc}
G & \longhookrightarrow & F & \longhookrightarrow & H & \longhookrightarrow & M \\
\updownarrow & & \updownarrow & & \updownarrow & & \\
G' & \longhookrightarrow & F' & \longhookrightarrow & H' & \longhookrightarrow & M'
\end{array}
$$

If the middle isomorphism is left out and H and H' are rela-beled F and F' respectively then we get what we want with both $m \in F$ and $m' \in F'$. □

5.12 Theorem: *(Assuming $V \cong On$)*

If M and M' are superuniversal then $M \cong M'$.

Proof: As $V \cong On$ there are enumerations $\{m_\alpha\}_{\alpha \in On}$ of M and $\{m'_\alpha\}_{\alpha \in On}$ of M'. By transfinite recursion on $\alpha \in On$ we will define $G_\alpha \trianglelefteq M$, $G'_\alpha \trianglelefteq M'$ and $i_\alpha : G_\alpha \cong G'_\alpha$ such that $m_\alpha \in G_\alpha$, $m'_\alpha \in G'_\alpha$ and whenever $\beta < \gamma$ then $G_\beta \trianglelefteq G_\gamma$, $G'_\beta \trianglelefteq G'_\gamma$ and the diagram

$$
\begin{array}{ccc}
G_\beta & \longhookrightarrow & G_\gamma \\
\updownarrow & & \updownarrow \\
G'_\beta & \longhookrightarrow & G'_\gamma
\end{array}
$$

commutes.

Once this is done it is clear that

$$
M = \bigcup_{\alpha \in On} G_\alpha, \ M' = \bigcup_{\alpha \in On} G'_\alpha \text{ and } i : M \cong M' \text{ where } i = \bigcup_{\alpha \in On} i_\alpha.
$$

So suppose that $G_\beta \trianglelefteq M$, $G'_\beta \trianglelefteq M'$ and $i_\beta : G_\beta \cong G'_\beta$ have been defined for $\beta < \alpha$ so that $m_\beta \in G_\beta$, $m'_\beta \in G'_\beta$ and whenever $\beta < \gamma$ the above conditions hold. Then $G \trianglelefteq M$, $G' \trianglelefteq M'$ and $i : G \cong G'$ where

$$
G = \bigcup_{\beta < \alpha} G_\beta, G' = \bigcup_{\beta < \alpha} G'_\beta \text{ and } i = \bigcup_{\beta < \alpha} i_\beta.
$$

So by the lemma we can choose $G_\alpha \trianglelefteq M$, $G'_\alpha \trianglelefteq M'$ and $i_\alpha : G_\alpha \cong G'_\alpha$ such that $G \trianglelefteq G_\alpha$, $G' \trianglelefteq G'_\alpha$, $m_\alpha \in G_\alpha$, $m'_\alpha \in G'_\alpha$ and the diagram

$$
\begin{array}{ccccc}
G & \longhookrightarrow & G_\alpha & \longhookrightarrow & M \\
\uparrow i & & \uparrow i_\alpha & & \\
G' & \longhookrightarrow & G'_\alpha & \longhookrightarrow & M'
\end{array}
$$

commutes. Note that a global form of the axiom of choice is needed to make the choice of G_α, G'_α and i_α for each $\alpha \in On$. But we do have this by the assumption that $V \cong On$.   □

Call a system M a GLOBALLY UNIVERSAL system if M is extensional and $M_0 \preceq M$ for each extensional-system M_0. Note that every globally universal system is locally universal.

5.13 Exercise: *Show that every superuniversal system is globally universal.*

5.14 Exercise: *Let BA_2 be the following axiom:*

If $f : (a, \in_a) \cong (b, \in_b)$, where a, b are transitive sets and $a' \supseteq a$ is also a transitive set then f has an extension to

$$f' : (a', \in_{a'}) \cong (b', \in_{b'})$$

for some transitive set $b' \supseteq b$.

Show that

(i) $BAFA \Longleftrightarrow BA_1 + BA_2$.

(ii) $FAFA_2 \Longrightarrow BA_2$.

The Existence of a Superuniversal System

We now turn to the construction of a superuniversal system. This will require the use of a quotient construction which always yields an extensional system in which a minimal number of identifications have been made. This is a dual construction to that using the maximal bisimulation $\equiv_M$ on a system M. Recall from chapter 2 that a binary relation R on M is a bisimulation on M if $R \subseteq R^+$, and that $\equiv_M$, the maximal bisimulation on M, is in fact an equivalence relation. We call R a CO-BISIMULATION relation if $R^+ \subseteq R$. The following exercise summarises the properties of the minimal reflexive bisimulation $\sim_M$ on M that we will need. (See also Hinnion 1981)

5.15 Exercise: *Show that*

(i) *There is a unique minimal reflexive co-bisimulation $\sim_M$ on a system M.*

(ii) *The relation $\sim_M$ is an equivalence relation on M that is also a bisimulation on M.*

(iii) *The system M is extensional iff*

$$a \sim_M b \implies a = b.$$

(iv) *If $\pi : M \longrightarrow M'$ is an injective system map then for $a, b \in M$*

$$a \sim_M b \iff \pi a \sim_{M'} \pi b,$$

5.16 Lemma: *(Assuming $V \cong On$) If M is a system then there is quotient $\pi : M \to M'$ of M with respect to $\sim_M$.*

Proof: As $V \cong On$ there is an enumeration $\{m_\alpha\}_{\alpha \in On}$ of M. For $a \in M$ let $\pi a = m_\alpha$ where α is the least ordinal such that $m_\alpha \sim_M a$. Then we clearly have

$$(*) \qquad a \sim_M b \iff \pi a = \pi b$$

for $a, b \in M$. Let M' be the system having as nodes the πa for $a \in M$ and having as edges $(\pi a, \pi b)$ for $a \longrightarrow b$ in M. As $\sim_M$ is a bisimulation M' is indeed a system and $\pi : M \longrightarrow M'$ is a surjective system map. It remains to show that M' is extensional. So let $(\pi a)_{M'} = (\pi b)_{M'}$. Then $\{\pi x \mid x \in a_M\} = \{\pi y \mid y \in b_M\}$ so that

$$\forall x \in a_M \exists y \in b_M (\pi x = \pi y) \quad \& \quad \forall y \in b_M \exists x \in a_M (\pi x = \pi y)$$

By $(*)$ and the definition of $\sim_M$ it follows that $a \sim_M b$ and hence $\pi a = \pi b$. □

Call a system map $\pi : M \longrightarrow M'$ given in this lemma a MINIMAL EXTENSIONAL QUOTIENT of M.

5.17 Exercise: *Let M be the system of extensional apgs. Let $\pi : M \longrightarrow M'$ be a minimal extensional quotient of M. Show that M' is globally universal.*

5.18 Theorem: *(Assuming $V \cong On$)*
There is a superuniversal system.

Proof: We shall give an inductive definition of a system M. The superuniversal system will be obtained as a minimal extensional quotient of M. The inductive definition will simultaneously generate the nodes of M and, as each node a of M is generated, it will also specify which previously generated nodes are to be the children of a.

Before giving the final definition of M we give an initial attempt and then improve it. So, as a first attempt, let M be the smallest system such that if

$$(*) \qquad G_0 \trianglelefteq G \text{ and } G_0 \trianglelefteq M$$

then for each $a \in G - G_0$

- $(G_0, G, a) \in M$
- $((G_0, G, a))_M \;=\; (a_G \cap G_0) \cup \{(G_0, G, x) \mid x \in a_G - G_0\}$ □

5.19 Exercise: *Verify that the inductive definition of M can be replaced by an explicit definition in ZFC^-.*

Observe that whenever G_0 and G satisfy $(*)$ then $\pi : G \longrightarrow M$ is a system map, where π is the extension of the identity map on G_0 such that for $a \in G - G_0$

$$\pi a \;=\; (G_0, G, a).$$

Thus $\pi : G \longrightarrow M$ completes the diagram

For the superuniversality of M we would want π to be injective, and it would be injective if $(G_0, G, x) \notin G_0$ for all $x \in G - G_0$. This is the case if the axiom of foundation holds. But without this assumption we appear to be stuck. For each i let $\sigma_i : V^3 \longrightarrow V$ be the injective map given by

$$\sigma_i(x, y, z) \;=\; (i, x, y, z) \text{ for all } x, y, z.$$

We now redefine M as follows:

Let M be the smallest system such that if

$$(*) \qquad G_0 \trianglelefteq G \text{ and } G_0 \trianglelefteq M$$

then for each $a \in G - G_0$ and each i

- $\sigma_i(G_0, G, a) \in M$

- $(\sigma_i(G_0, G, a))_M = (a_G \cap G_0) \cup \{\sigma_i(G_0, G, x) \mid x \in a_G - G_0\}$

Now if we are given G_0 and G satisfying $(*)$ then for each $i \in I$ the map $\pi_i : G \longrightarrow M$ is a system map extending the identity map on G_0 such that for $a \in G - G_0$

$$\pi_i a = \sigma_i(G_0, G, a)$$

Now the map π_i will be injective provided that $\sigma_i(G_0, G, x) \notin G_0$ for all $x \in G - G_0$. As G_0 is small this must be the case for some choice of i. For otherwise there would be an injective map assigning $\sigma_i(G_0, G, x_i) \in G_0$ for some $x_i \in G - G_0$ to any i. Hence we can complete the diagram

with an injective system map $\pi_i : G \longrightarrow M$.

A superuniversal system has to be extensional and M is not. So let us form a minimal extensional quotient $\pi : M \longrightarrow M'$ of M. We will show that M' is superuniversal. M' is extensional by definition. Let $G_0' \trianglelefteq M'$ and $G_0' \trianglelefteq G'$, where G' is an extensional graph. We must find an injective system map $\pi' : G' \longrightarrow M'$ that extends the identity map on G_0'. As $\pi : M \longrightarrow M'$ is surjective and $G_0' \trianglelefteq M'$ we may use the axiom of choice to find $G_0 \trianglelefteq M$ such that when π is restricted to G_0 it is a surjection $\psi_0 : G_0 \longrightarrow G_0'$. Find an injective function σ defined on $G' - G_0'$ so that $\sigma x \notin G_0$ for $x \in G' - G_0'$. Now let G be the graph having as nodes the nodes of G_0 and the σx. If $x \in G_0$ then let $x_G = x_{G_0}$ so that $G_0 \trianglelefteq G$. If $x \in G' - G_0'$ let

$$(\sigma x)_G = \{\sigma y \mid y \in x_{G'} \ \& \ y \notin G_0'\} \ \cup \ \{z \in G_0 \mid \psi_0 z \in x_{G'}\}.$$

Define $\psi : G \longrightarrow G'$ so that it extends $\psi_0 : G_0 \longrightarrow G_0'$ and so that $\psi(\sigma x) = x$ for $x \in G' - G_0'$. So we get the commutative diagram

$$
\begin{array}{ccc}
G & \xrightarrow{\ \psi\ } & G' \\
\uparrow & & \uparrow \\
G_0 & \xrightarrow{\ \psi_0\ } & G_0'
\end{array}
$$

Now G_0 and G satisfy $(*)$ so that by our earlier work we can find an injective system map $\pi_i : G \longrightarrow M$ extending the identity map on G_0. We now have the commutative diagram

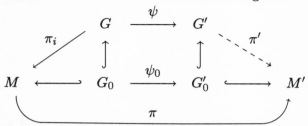

which we wish to complete with an injective system map $\pi' : G' \longrightarrow M'$ extending the identity map on G_0'. We need the following result.

5.20 Lemma: *For $x, y \in G$*

$$\psi x = \psi y \iff \pi(\pi_i x) = \pi(\pi_i y).$$

Proof: Observe that by part (iv) of exercise 5.15.

$$\pi(\pi_i x) = \pi(\pi_i y) \iff \pi_i x \sim_M \pi_i y \iff x \sim_G y.$$

For $x, y \in G$ let

$$x R y \iff \psi x = \psi y.$$

We must show that $x R y \iff x \sim_G y$. If $x R^+ y$ then

$$\{\psi x' \mid x' \in x_G\} = \{\psi y' \mid y' \in y_G\}$$

so that $(\psi x)_{G'} = (\psi y)_{G'}$ and hence $\psi x = \psi y$, as G' is extensional. So $R^+ \subseteq R$. As R is reflexive it follows that $x \sim_G y \implies x R y$. For the converse implication let $x R y$; i.e. $\psi x = \psi y$. If $\psi x \in G_0'$ then $x, y \in G_0$ and $\psi_0 x = \psi_0 y$ so that $\pi x = \pi y$ and hence $x \sim_M y$. It follows that $x \sim_{G_0} y$ and hence $x \sim_G y$. Here we have made two further applications of part (iv) of exercise 5.15. If $\psi x \in G' - G_0'$ then $x = y$ so that, as $\sim_G$ is reflexive, $x \sim_G y$. $\qquad\square$

By this lemma and the fact that $\psi : G \longrightarrow G'$ is surjective there is a unique injective map $\pi' : G' \longrightarrow M'$ such that for $x \in G$

$$\pi'(\psi x) = \pi(\pi_i x).$$

It is easy to check that π is a system map that extends the identity map on G_0'. $\qquad\square$

Part Three

On Using the Anti-Foundation Axiom

6 | Fixed Points of Set Continuous Operators

In this chapter we consider the notion of a set continuous operator. Each such operator will be shown to have both a least and a greatest fixed point.

Set Continuous Operators

6.1 Definition: *Let Φ be a class operator; i.e. ΦX is a class for each class X. Φ is* SET CONTINUOUS *if for each class X*

$$\Phi X \;\; = \;\; \bigcup \{ \Phi x \mid x \in V \;\&\; x \subseteq X \}.$$

This is equivalent to the conjunction of the following two conditions.

(1) *Φ is* MONOTONE; *i.e.*

$$X \subseteq Y \;\;\Longrightarrow\;\; \Phi X \subseteq \Phi Y.$$

(2) *Φ is* SET BASED; *i.e.*

$$a \in \Phi X \;\;\Longrightarrow\;\; a \in \Phi x \text{ for some set } x \subseteq X.$$

Set continuity has alternative characterizations given by the following exercise.

6.2 Exercise: *Let Φ be a class operator. Show that the following are equivalent.*

(i) *Φ is set continuous.*

(ii) *There is a class relation R such that for all classes X*

$$\Phi X \;\; = \;\; \{ a \mid \exists x \in V \; aRx \;\&\; x \subseteq X \}.$$

(iii) *There is a map $\nu : \Delta \to V$, for some class Δ, and a family $(\tau_\delta)_{\delta \in \Delta}$ of maps $\tau_\delta : V^{\nu \delta} \to V$ such that for all classes X*

$$\Phi X \;=\; \{\tau_\delta f \mid \delta \in \Delta \;\&\; f \in X^{\nu \delta}\}.$$

Obvious examples of set continuous operators are *pow*, *Id* and K_A for each class A, where for each class X

$$\begin{aligned} pow\, X &= \{x \in V \mid x \subseteq X\}, \\ Id\, X &= X, \\ K_A\, X &= A. \end{aligned}$$

Also the composition $\Phi \circ \Psi$ of two set continuous operators Φ, Ψ is clearly set continuous.

For any system M we have the following example of a set continuous operator used in the construction of the maximal bisimulation on M. For each class X $\Phi_M X$ is the class of pairs $(a, b) \in M \times M$ such that

$$\forall x \in a_M \; \exists y \in b_M \; (x, y) \in X \;\;\&\;\; \forall y \in b_M \; \exists x \in a_M \; (x, y) \in X$$

The next exercise details some ways of constructing new set continuous operators out of old ones.

6.3 Exercise: *Let $(\Phi_i)_{i \in I}$ be a family of set continuous operators indexed by the class I.*

(i) *Show that $\sum_{i \in I} \Phi_i$ is set continuous, where for each class X*

$$\Big(\sum_{i \in I} \Phi_i\Big) X \;=\; \sum_{i \in I} \Phi_i X.$$

(ii) *If I is a set show that $\prod_{i \in I} \Phi_i$ is set continuous, where for each class X*

$$\Big(\prod_{i \in I} \Phi_i\Big) X \;=\; \prod_{i \in I} (\Phi_i X).$$

(iii) *If $I = \{1, \ldots n\}$ show that $\Phi_1 \times \cdots \times \Phi_n$ is set continuous, where for each class X*

$$(\Phi_1 \times \cdots \times \Phi_n) X \;=\; (\Phi_1 X) \times \cdots \times (\Phi_n X).$$

Note that for set continuous $\Phi_1, \ldots \Phi_n$ we can also define the set continuous operator $\Phi_1 + \cdots + \Phi_n$, where

$$\Phi_1 + \cdots + \Phi_n \; = \; \sum_{i \in I} \Phi_i$$

when $I = \{1, \ldots, n\}$. Also if Φ is set continuous then so is Φ^I for each set I, where $\Phi^I = \prod_{i \in I} \Phi_i$ when $\Phi_i = \Phi$ for each $i \in I$.

Using the results of this exercise a great variety of set continuous operators can be formed. An example, chosen more or less at random, is the set continuous operator $\Phi = (pow((pow\,Id) + Id^I)) \times K_A$, where I is a set and A is a class. This is the operator such that for each class X

$$\Phi X \quad = \quad pow(pow\,X + X^I) \times A$$

for all classes X.

Fixed Points

We now turn to the construction of the least and greatest fixed points of a set continuous operator. If Φ is a set continuous operator let $I_\Phi \;\; = \;\; \{fi \mid (f, \prec, i) \in B\}$ where B is the class of triples $(f, \prec, i)$ such that f is a function, $\prec$ is a well-founded relation on the set $\mathrm{dom}f$, $i \in \mathrm{dom}f$ and for all $j \in \mathrm{dom}f$

$$fj \in \Phi\{fk \mid k \prec j\}.$$

6.4 Theorem: *If Φ is a set continuous operator and $I = I_\Phi$ then*

(1) $\Phi I \subseteq I$,

(2) *If $\Phi X \subseteq X$ then $I \subseteq X$,*

(3) *I is the least fixed point of Φ.*

Proof:

(1) Let $a \in \Phi I$. Then, as Φ is set based, there is a set x such that $a \in \Phi x$ and $x \subseteq I$, so that

$$\forall y \in x \;\; \exists (f, \prec, i) \in B \;\; y = fi.$$

By the collection scheme there is a set $A_0 \subseteq B$ such that

$$\forall y \in x \;\; \exists (f, \prec, i) \in A_0 \;\; y = fi.$$

Let $A \;\; = \;\; A_0 \cup \{*\}$ where $* \notin A_0$. Let $\prec\!\!\prec$ be the least relation on A such that for all $u \in A_0$ $u \prec\!\!\prec *$ and whenever $(f, \prec, i) \in A_0$ and $i \prec j$ then $(f, \prec, i) \prec\!\!\prec (f, \prec, j)$. Then $\prec\!\!\prec$ is clearly well-founded and we can define the following function F with domain A.

$$F* \;\; = \;\; a,$$

$$F(f, \prec, i) \ = \ fi$$

for $(f, \prec, i) \in A_0$. Observe that $(F, \prec\!\!\prec, *) \in B$ so that, as $a = F*$, $a \in I$.

(2) Let $\Phi X \subseteq X$ and let $a \in I$. We must show that $a \in X$. There is $(f, \prec, i) \in B$ such that $a = fi$. It suffices to show that $fj \in X$ for all $j \in \mathrm{dom} f$. We do this by induction on the well-founded relation $\prec$. So suppose that $fk \in X$ for all $k \prec j$. Then $\{fk \mid k \prec j\} \subseteq X$ so that, as $fj \in \Phi\{fk \mid k \prec j\}$, $fj \in \Phi X$.

(3) By (1) and the monotonicity of Φ

$$\Phi(\Phi I) \subseteq \Phi I.$$

Hence by (2) $I \subseteq \Phi I$. This and (1) imply that I is a fixed point of Φ. By (2) it must be the least fixed point of Φ. $\square$

If Φ is a set continuous operator let $J_\Phi = \bigcup\{x \in V \mid x \subseteq \Phi x\}$.

6.5 Theorem: *If Φ is a set continuous operator and $J = J_\Phi$ then*

(1) $J \subseteq \Phi J$,
(2) *If $X \subseteq \Phi X$ then $X \subseteq J$,*
(3) *J is the largest fixed point of Φ.*

Proof:

(1) Let $a \in J$. Then $a \in x$ for some set x such that $x \subseteq \Phi x$. It follows that $a \in \Phi J$ as $x \subseteq J$ and Φ is monotone.

(2) Let $X \subseteq \Phi X$ and let $a \in X$. We must show that $a \in J$. We first show that for each set $x \subseteq X$ there is a set $x' \subseteq X$ such that $x \subseteq \Phi x'$. So let $x \subseteq X$. Then $x \subseteq \Phi X$ so that

$$\forall y \in x \ \exists u \ y \in \Phi u \ \& \ u \subseteq X.$$

By the collection axiom scheme there is a set A such that

$$\forall y \in x \ \exists u \in A \ y \in \Phi u \ \& \ u \subseteq X.$$

If we let $x' = \bigcup\{u \in A \mid u \subseteq X\}$ then x' is a subset of X and $x \subseteq \Phi x'$ as required.

Now we can use the axiom of dependent choices to find an infinite sequence $x_0, x_1, \ldots$ of subsets of X such that

$x_0 = \{a\}$ and $x_n \subseteq \Phi x_{n+1}$ for all n. Let $x = \bigcup_n x_n$. Then x is a set and if $y \in x$ then $y \in x_n$ for some n so that $y \in x_n \subseteq \Phi x_{n+1} \subseteq \Phi x$. Thus $x \subseteq \Phi x$. As $a \in x_0 \subseteq x$ it follows that $a \in J$. I do not know if this use of the axiom of dependent choices was essential.

(3) The argument to show that J is the largest fixed point of Φ is simply dual to the argument, at the end of the proof of the previous theorem, that I is the least fixed point.   □

In certain cases the fixed points I_Φ and J_Φ of a set continuous operator Φ are equal. In these cases I_Φ is the unique fixed point of Φ. For example if we assume the axiom of foundation then V is the unique fixed point of pow and $\emptyset$ is the unique fixed point of Φ where $\Phi X = A \times X$ for all classes X. Of course when AFA is assumed pow and Φ have many fixed points. Recall that $I_{pow} = V_{wf}$ while $J_{pow} = V$. Also $I_\Phi = \emptyset$ while J_Φ is the class of all streams $(a_0, (a_1, (a_2, \ldots)))$ of elements $a_0, a_1, a_2, \ldots$ of A.

The following gives a sufficient condition for a set continuous operator to have a unique fixed point.

6.6 Exercise: *Let Φ be a set continuous operator such that there is a well-founded class relation $\prec$ such that for all classes X and all $a \in \Phi X$*

$$a \in \Phi\{x \in X \mid x \prec a\}.$$

Show that $I_\Phi = J_\Phi$.

There is a standard approach to finding fixed points of operators by using transfinite recursion to define iterations of the operator. But the definition of transfinite sequences of classes by transfinite recursion requires strong impredicative comprehension principles for defining classes. As these are not available in ZFC^- we have not used this approach to define I_Φ and J_Φ. But once those classes have been defined the iterations of Φ can be obtained in ZFC^- as spelled out in the following exercise.

6.7 Exercise: *Let Φ be set continuous. Working in ZFC^- show that there are classes I^α and J^α, for $\alpha \in On$, so that*

$$I^\alpha = \Phi(\bigcup_{\beta < \alpha} I^\beta),$$

$$J^\alpha \;=\; \Phi(\bigcap_{\beta<\alpha} J^\beta).$$

Show also that

$$I_\Phi \;=\; \bigcup_{\alpha\in On} I^\alpha,$$

$$J_\Phi \;=\; \bigcap_{\alpha\in On} J^\alpha.$$

Often a set continuous operator has the following additional property.

6.8 Definition: *The class operator* Φ PRESERVES INTERSECTIONS *if for every family of classes* $(X_i)_{i\in I}$

$$\Phi(\bigcap_{i\in I} X_i) \;=\; \bigcap_{i\in I} \Phi X_i.$$

If the set continuous operator Φ does preserve intersections then $\Phi(\bigcap_{n<\omega} J^n) \;=\; \bigcap_{n<\omega} \Phi J^n \;=\; \bigcap_{n<\omega} J^n$. It follows that this is the largest fixed point J_Φ.

6.9 Exercise: *Show that:*

(i) *If* Φ *is defined as in (ii) of exercise 6.2. and for all* a, x, y

$$aRx \;\&\; aRy \;\implies\; x = y$$

then Φ *preserves intersections.*

(ii) *If* Φ *is defined as in (iii) of exercise 6.2. and for all* $\delta_1, \delta_2 \in \Delta$ *and all* $f_1 : \nu\delta_1 \to V, f_2 : \nu\delta_2 \to V$

$$\tau_{\delta_1} f_1 \;=\; \tau_{\delta_2} f_2 \;\implies\; \mathrm{ran} f_1 \;=\; \mathrm{ran} f_2$$

then Φ *preserves intersections.*

We end this chapter with a useful application of *AFA*. We use the terminology of the Substitution and Solution lemmas of chapter 1. So let X be a class of atoms and let Φ be a set continuous operator with largest fixed point J. We call an X-set a a Φ-LOCAL set if for every class B of pure sets and every $\tau : X \to B$

$$\hat\tau a \in \Phi B.$$

6.10 Theorem: *(assuming AFA) Let a_x be a Φ-local X-set for each atom x in X. Let $\pi = (b_x)_{x \in X}$ be the unique solution, which exists by the solution lemma, of the system of equations*

$$x = a_x \quad (x \in X).$$

Then $b_x \in J$ for all $x \in X$.

Proof: Let $B = \{b_x \mid x \in X\}$. If $b \in B$ then $b = b_x = \hat{\pi} a_x$ for some $x \in X$ so that, as a_x is Φ-local, $b \in \Phi B$. Thus $B \subseteq \Phi B$ so that $B \subseteq J$. $\square$

As an example of the use of this result let $\Phi X = A \times X$ for each class X, where A is some fixed class. Let $a_0, a_1, \ldots \in A$ and let $(b_n)_{n=0,1,\ldots}$ be the solution to the system of equations

$$x_n = (a_n, x_{n+1}) \quad (n = 0, 1, \ldots).$$

As (a_n, x_{n+1}) is Φ-local for each n it follows that $b_n \in J$ for each n.

7 | The Special Final Coalgebra Theorem

Perhaps the main result of Part 1 was Theorem 3.10. That theorem with theorem 3.8 characterize the full models of AFA as those systems M such that for every system M' there is a unique system map $M' \to M$. Assuming AFA, the largest fixed point V of pow is such a system M. The aim of this chapter is to give a generalization of this result. To do so we need to make use of some notions from category theory and view pow as a functor on the (superlarge) category of classes and maps between classes. This is simply done. If $\pi : A \to B$ is a map then $pow\ \pi : pow\ A \to pow\ B$ is defined by

$$(pow\ \pi)x = \{\pi a \mid a \in x\}$$

for all $x \in pow A$.

A system may be viewed as consisting of a class M of nodes and a map $()_M : M \to pow M$, where for each $a \in M$ a_M is the set of children of a. A system map from M to another system M' is a map $\pi : M \to M'$ such that for all $a \in M$

$$(\pi a)_{M'} = (pow\ \pi)a_M;$$

i.e., such that the following diagram commutes.

$$
\begin{array}{ccc}
M & \xrightarrow{\ \ \pi\ \ } & M' \\
{\scriptstyle ()_M}\big\downarrow & & \big\downarrow{\scriptstyle ()_{M'}} \\
pow\ M & \xrightarrow[pow\ \pi]{} & pow\ M'
\end{array}
$$

When the notions of system and system map are viewed in this way the desired generalization becomes clear. Systems are simply coalgebras, in the sense defined below, for the functor *pow*, and the system maps are the coalgebra homomorphisms. The notion of coalgebra will be defined as a dual to the more familiar notion of an algebra.

Initial Algebras and Final Coalgebras

We start with a formulation of the general notions we will be using. We assume given a fixed functor $\Phi : \mathcal{C} \to \mathcal{C}$ where $\mathcal{C}$ is a category.

The following notions are relative to this functor.

7.1 Definition:

(1) (A, α) is an ALGEBRA if $\alpha : \Phi A \to A$ in $\mathcal{C}$.
When α is understood we shall just use A for the algebra.
If α is a bijection then the algebra is a FULL ALGEBRA.

(2) Given algebras (A, α) and (B, β) π is a HOMOMORPHISM from (A, α) to (B, β), written $\pi : (A, \alpha) \to (B, \beta)$, if $\pi : A \to B$ such that the diagram

$$
\begin{array}{ccc}
\Phi A & \xrightarrow{\ \Phi\pi\ } & \Phi B \\
\alpha \downarrow & & \downarrow \beta \\
A & \xrightarrow{\ \pi\ } & B
\end{array}
$$

commutes.
Algebras and homomorphisms form a category. The general notion of an initial object when applied to the category of algebras gives us

(3) (A, α) is an INITIAL ALGEBRA if it is an algebra such that for every algebra (B, β) there is a unique homomorphism $(A, \alpha) \to (B, \beta)$.

7.2 Exercise: Show that

(i) Any two initial algebras are isomorphic.

(ii) Any initial algebra is full.

The functor Φ may be viewed as a functor Φ^{op} on the category $\mathcal{C}^{op}$ dual to $\mathcal{C}$. $\mathcal{C}^{op}$ has the same objects as $\mathcal{C}$, but a map $f : A \to B$ in $\mathcal{C}$ is viewed as a map $f : B \to A$ in $\mathcal{C}^{op}$.

We call (A, α) a COALGEBRA (relative to Φ) if it is an algebra relative to Φ^{op}. Moreover (A, α) is a final coalgebra (relative to Φ)

if it is an initial algebra relative to Φ^{op}. Thus (A, α) is a final coalgebra if $\alpha : A \to \Phi A$ in $\mathcal{C}$ such that whenever $\beta : B \to \Phi B$ in $\mathcal{C}$ there is a unique map $\pi : B \to A$ in $\mathcal{C}$ such that the diagram

$$
\begin{array}{ccc}
\Phi A & \xleftarrow{\;\;\Phi\pi\;\;} & \Phi B \\
\alpha \uparrow & & \uparrow \beta \\
A & \xleftarrow[\pi]{} & B
\end{array}
$$

commutes.

Thus the notion of final coalgebra is dual to that of initial algebra and the results of the exercise give dual results for final coalgebras.

Standard Functors

From now on we fix $\mathcal{C}$ to be the superlarge category whose objects are classes and whose maps are the class maps between classes. The excessive size of this category is not a serious problem. It can be overcome in a straightforward way. But the details would be out of place here.

7.3 Definition: *A functor $\Phi : \mathcal{C} \to \mathcal{C}$ is* STANDARD *if it is set continuous as a class operator and preserves inclusion maps; i.e., if $X \subseteq Y$ then $\Phi i_{X,Y} = i_{\Phi X, \Phi Y}$ where $i_{X,Y} : X \to Y$ is the inclusion map.*

An example of a standard functor is the functor *pow* defined at the start of this section. Trivially the identity functor Id and the constant functors K_A, for each class A, are standard. Also it is clear that the composition $\Phi \circ \Psi$ of standard functors Φ and Ψ is a standard functor. The following exercise gives further ways to construct new standard functors from old ones.

7.4 Exercise: *Let $(\Phi_i)_{i \in I}$ be a family of standard functors indexed by the class I.*

(i) *Show that $\sum_{i \in I} \Phi_i$ is a standard functor Φ where if X is a class*

$$\Phi X = \sum_{i \in I} \Phi_i X,$$

and

$$(\Phi \pi)(i, a) = (i, (\Phi_i \pi)a)$$

if $\pi : X \to Y$ and $(i, a) \in \Phi X$.

(ii) *Show that if I is a set then $\prod_{i \in I} \Phi_i$ is a standard functor Φ where if X is a class*

$$\Phi X = \prod_{i \in I} \Phi_i X$$

and

$$((\Phi \pi) f) i = (\Phi_i \pi)(f i)$$

if $\pi : X \to Y$, $f \in \Phi X$ and $i \in I$.

(iii) *Show that if $I = \{i, \ldots, n\}$ then $\Phi_1 \times \cdots \times \Phi_n$ is a standard functor Φ where if X is a class*

$$\Phi X = \Phi_1 X \times \cdots \times \Phi_n X$$

and

$$(\Phi \pi)(a_1, \ldots, a_n) = ((\Phi_1 \pi) a_1, \ldots, (\Phi_n \pi) a_n)$$

if $\pi : X \to Y$ and $a_i \in \Phi_i X$ for $i = 1, \ldots, n$.

7.5 Exercise: *Assume that the set continuous operator Φ is defined in terms of a family $(\tau_\delta)_{\delta \in \Delta}$ as in part (iii) of exercise 6.2; i.e. for each class X*

$$\Phi X \;=\; \{\tau_\delta f \mid \delta \in \Delta \ \& \ f \in X^{\nu \delta}\}.$$

Assume that

$$\tau_{\delta_1} f_1 = \tau_{\delta_2} f_2 \implies \tau_{\delta_1}(\pi \circ f_1) = \tau_{\delta_2}(\pi \circ f_2)$$

whenever $\pi : X \to Y, \delta_1, \delta_2 \in \Delta$ and $f_1 \in X^{\nu \delta_1}, f_2 \in X^{\nu \delta_2}$.
Show that Φ can be made a standard functor by defining

$$(\Phi \pi)(\tau_\delta f) = \tau_\delta(\pi \circ f)$$

whenever $\pi : X \to Y, \delta \in \Delta$ and $f \in X^{\nu \delta}$.

We shall need to use the following property of standard functors. If $\pi : X \to Y$ and $Z \subseteq X$ then

$$\Phi(\pi \restriction Z) = (\Phi \pi) \restriction (\Phi Z).$$

This is because $\pi \restriction Z = \pi \circ i_{Z,X}$.

Also note that any fixed point of a standard functor can be viewed as a full algebra or full coalgebra, using the identity map. We have the following result concerning the least fixed point.

7.6 Theorem:

If Φ is a standard functor then I_Φ is an initial algebra.

Proof: Let (A, α) be an algebra. We must show that there is a unique homomorphism from I_Φ to (A, α); i.e., a map $\pi : I_\Phi \to A$ such that for all $x \in I_\Phi$

$$\pi x = \alpha(\ (\Phi\pi)\ x).$$

Recall that $I_\Phi = \bigcup_{\lambda \in On} I^\lambda$, where $I^\lambda = \Phi(\bigcup_{\mu < \lambda} I^\mu)$. (See exercise 6.7) Note that $I^\mu \subseteq I^\lambda$ for $\mu < \lambda$. We will define $\pi^\lambda : I^\lambda \to A$, by transfinite recursion on $\lambda \in On$, so that for $\mu < \lambda$

$$\pi^\mu = \pi^\lambda \restriction I^\mu.$$

The problem of checking that this definition can be carried out in ZFC^- will be ignored here. The desired map π will be defined as the union of the π^λ.

As induction hypothesis we assume that $\pi^\mu : I^\mu \to A$ has been defined for all $\mu < \lambda$ so that

$$\pi^\nu = \pi^\mu \restriction I^\nu \text{ for } \nu < \mu < \lambda.$$

Let $I^{<\lambda} = \bigcup_{\mu < \lambda} I^\mu$ and $\pi^{<\lambda} = \bigcup_{\mu < \lambda} \pi^\mu$. By our induction hypothesis $\pi^{<\lambda}$ is a well-defined map $I^{<\lambda} \to A$ such that

$$\pi^\mu = \pi^{<\lambda} \restriction I^\mu \text{ for } \mu < \lambda$$

It follows that $\Phi\pi^{<\lambda} : I^\lambda \to \Phi A$ so that we can define

$$\pi^\lambda = \alpha \circ (\Phi\pi^{<\lambda}).$$

We will assume that the π^μ, for $\mu < \lambda$, have also been defined in that way so that

$$\pi^\mu = \alpha \circ (\Phi\pi^{<\mu}) \text{ for } \mu < \lambda.$$

As $\pi^{<\mu} = \pi^{<\lambda} \restriction I^{<\mu}$ for $\mu < \lambda$, we get

$$\begin{aligned}
\pi^\mu &= \alpha \circ (\Phi(\pi^{<\lambda} \restriction I^{<\mu})) \\
&= \alpha \circ ((\Phi\pi^{<\lambda}) \restriction (\Phi I^{<\mu})) \\
&= \pi^\lambda \restriction I^\mu,
\end{aligned}$$

as required.

Having defined π^λ for $\lambda \in On$ we can define $\pi : I_\Phi \to A$ to be the union of the π^λ. Then $\pi^\lambda = \pi \restriction I^\lambda$ and $\pi^{<\lambda} = \pi \restriction I^{<\lambda}$ for $\lambda \in On$. So, if $x \in I_\Phi$ then $x \in I^\lambda$ for some $\lambda \in On$, so that

$$\begin{aligned}
\pi x &= \pi^\lambda x \\
&= \alpha((\Phi\pi^{<\lambda})x) \\
&= \alpha((\Phi(\pi \restriction I^{<\lambda}))x) \\
&= \alpha(((\Phi\pi) \restriction I^\lambda)x) \\
&= \alpha((\Phi\pi)x).
\end{aligned}$$

This shows the existence of π. For uniqueness, suppose that $\tau : I_\Phi \to A$ such that for $x \in I_\Phi$

$$\tau x = \alpha((\Phi\tau)x).$$

Then I claim that $\tau \restriction I^\lambda = \pi^\lambda$ for all $\lambda \in On$ so that $\tau = \pi$. For if $\tau \restriction I^\mu = \pi^\mu$ for all $\mu < \lambda$ then $\tau \restriction I^{<\lambda} = \pi^{<\lambda}$ so that if $x \in I^\lambda$ then

$$\begin{aligned}
\tau x &= \alpha((\Phi\tau)x) \\
&= \alpha(((\Phi\tau) \restriction I^\lambda)x) \\
&= \alpha((\Phi(\tau \restriction I^{<\lambda}))x) \\
&= \alpha((\Phi(\pi^{<\lambda}))x) \\
&= \pi^\lambda x.
\end{aligned}$$

Thus $\tau \restriction I^\lambda = \pi^\lambda$. $\qquad\qquad\qquad\qquad\qquad\qquad\qquad\qquad$ $\square$

The Final Coalgebra Theorems

It is natural to consider the dual to the previous theorem. But such a dual is somewhat harder to come by. To see this observe that if we assume the axiom of foundation then V is the unique fixed point of the standard functor *pow*. But the class of well-founded sets certainly does not give a final coalgebra for *pow*, as by theorem 3.8 any final coalgebra for *pow* is a model of the anti-foundation axiom. On the other hand if we assume the anti-foundation axiom then the largest fixed point V of *pow* is indeed a final coalgebra.

The following result does give the existence of final coalgebras. A WEAK PULLBACK is a commutative square

$$X_0 \xrightarrow{\;\; p_1 \;\;} X_1$$

$$p_2 \downarrow \qquad\qquad \downarrow q_1$$

$$X_2 \xrightarrow[\;\; q_2 \;\;]{} Y$$

such that for all $x_1 \in X_1$, $x_2 \in X_2$ such that $q_1 x_1 = q_2 x_2$ there is $x \in X_0$ such that

$$x_1 = p_1 x \text{ and } x_2 = p_2 x.$$

Final Coalgebra Theorem:
Any standard functor that preserves weak pullbacks has a final coalgebra.

We will outline a proof of the final coalgebra theorem. The construction of a final coalgebra will generalise the construction of V_c in chapter 3. We assume given a fixed standard functor Φ. A coalgebra (X, α) is a COMPLETE coalgebra if for every small coalgebra (Y, β) there is a unique homomorphism $(Y, \beta) \to (X, \alpha)$. It is not hard to show that a coalgebra is final if it is complete. The unique homomorphism from a possibly large coalgebra (Y, β) to a complete coalgebra is obtained by piecing together the unique homomorphisms from the small subcoalgebras of (Y, β) to the complete coalgebra.

A coalgebra (X, α) is a WEAKLY COMPLETE coalgebra [STRONGLY EXTENSIONAL coalgebra] if for every small coalgebra (Y, β) there is at least one [at most one] homomorphism $(Y, \beta) \to (X, \alpha)$. Obviously a coalgebra is complete if and only if it is both weakly complete and strongly extensional. It is easy to construct a weakly complete coalgebra (C, γ). Let C be the class of pointed small coalgebras; i.e. triples (X, α, x) where (X, α) is a small coalgebra and $x \in X$. Now define $\gamma : C \to \Phi C$ as follows. First let $\alpha^* : X \to C$ be given by

$$\alpha^* x \;=\; (X, \alpha, x)$$

for all $x \in X$. Define

$$\gamma(X, \alpha, x) \;=\; (\Phi \alpha^*)(\alpha x)$$

for $(X, \alpha, x) \in C$. To see that the coalgebra (C, γ) is weakly complete observe that if (X, α) is a small coalgebra then $\alpha^* : (X, \alpha) \to (C, \gamma)$ is a homomorphism.

The following result is the key to the construction of a complete coalgebra from a weakly complete one.

7.7 Lemma: *If Φ preserves weak pullbacks then for each coalgebra (X, α) there is a strongly extensional coalgebra $(\overline{X}, \overline{\alpha})$ and a surjective homomorphism $(X, \alpha) \to (\overline{X}, \overline{\alpha})$.*

If we apply this lemma to the weakly complete coalgebra (C, γ) then we get a strongly extensional coalgebra $(\overline{C}, \overline{\gamma})$. Because of the homomorphism $(C, \gamma) \to (\overline{C}, \overline{\gamma})$ the weak completeness of (C, γ) carries over trivially to $(\overline{C}, \overline{\gamma})$ so that $(\overline{C}, \overline{\gamma})$ is both strongly extensional and weakly complete and so is complete and therefore final.

The lemma will not be proved in general, but we will outline a proof for the special case of the functor Φ where

$$\Phi = pow \circ (K_A \times Id),$$

where A is some fixed class. In this case a coalgebra for Φ has the form (X, α) where X is a class and $\alpha : X \to pow(A \times X)$. Such a coalgebra determines a system (X, α_a) for each $a \in A$, where $\alpha_a : X \to pow X$ is given by

$$\alpha_a x = \{y \in X \mid (a, y) \in x\}$$

for each $x \in X$. If $R \subseteq X \times X$ is a bisimulation relation on (X, α_a) for each $a \in A$ then call R a bisimulation relation on the coalgebra (X, α). As with maximal bisimulations on systems, it is not difficult to show that every coalgebra (X, α) has a maximal bisimulation and moreover that the relation is an equivalence relation. The next step is to form a quotient $\pi : X \to \overline{X}$ of the class X with respect to this equivalence relation. By suitably defining $\overline{\alpha} : \overline{X} \to \Phi\overline{X}$ we can get a coalgebra $(\overline{X}, \overline{\alpha})$ so that π is a surjective homomorphism $(X, \alpha) \to (\overline{X}, \overline{\alpha})$. Finally it is necessary to show that $(\overline{X}, \overline{\alpha})$ is strongly extensional, but that is straightforward.

To get a better dual to the initial algebra theorem we will need to assume *AFA* and replace the condition on the standard functor of preserving weak pullbacks by a seemingly quite different condition. In order to formulate this new condition we will need to use the expanded universe of sets involved in the solution lemma in chapter 1. Recall that the expanded universe has an atom x_i for each pure set i. If x is such an atom let i_x be the pure set i such that $x = x_i$. Given a class A of pure sets let

$$X_A = \{x_i \mid i \in A\},$$

and if $\pi : X_A \to V$ let $\pi' : A \to V$ be given by

$$\pi' i = \pi x_i \text{ for all } i \in A.$$

A standard functor Φ is defined to be UNIFORM ON MAPS if for each class A of pure sets there is a family $(c_u)_{u \in \Phi A}$, where c_u is an X_A-set for each $u \in \Phi A$, such that for all $\pi : X_A \to V$ and all $u \in \Phi A$

$$(\Phi \pi') u = \hat{\pi} c_u.$$

The Special Final Coalgebra Theorem:
(Assuming AFA) If Φ is a standard functor that is uniform on maps then J_ϕ is a final coalgebra.

Proof: Let (A, α) be a coalgebra for Φ. So $\alpha : A \to \Phi A$. Let c_u be an X_A-set for each $u \in \Phi A$ such that for all $\pi : X_A \to V$ and all $u \in \Phi A$

$$(\Phi \pi') u = \hat{\pi} c_u.$$

For each $x \in X_A$ let a_x be the X_A-set $c_{\alpha i_x}$.

Note that each X_A-set a_x is Φ-local. For if B is a class of pure sets and $\tau : X_A \to B$ then

$$\hat{\tau} a_x = \hat{\tau} c_{\alpha i_x} = (\Phi \tau')(\alpha i_x)$$

so that, as $\alpha i_x \in \Phi A$ and $\Phi \tau' : \Phi A \to \Phi B$, it follows that $\hat{\tau} a_x \in \Phi B$.

By the solution lemma the system of equations

$$x = a_x \quad (x \in X_A)$$

has a unique solution, and by theorem 6.10. that solution is a map $\pi : X_A \to J_\Phi$. It follows that $\pi' : A \to J_\Phi$ such that for all $i \in A$

$$\pi' i = \pi x_i = \hat{\pi} a_{x_i} = \hat{\pi} c_{\alpha i} = (\Phi \pi')(\alpha i),$$

so that the diagram

$$A \xrightarrow{\ \pi'\ } J_\Phi$$

$$\alpha \downarrow \qquad\qquad \|$$

$$\Phi A \xrightarrow[\Phi\pi']{} \Phi J_\Phi$$

commutes. This means that π' is a homomorphism from the coalgebra (A, α) to the coalgebra J_Φ. As the solution π is unique, it follows that the homomorphism π' is unique. □

In practice the natural functors always seem to be uniform on maps. For example to see that *pow* is uniform on maps let $b_u = \{2\} \times \{x_i \mid i \in u\}$ for $u \in pow\ I$. It is also the case that the natural standard functors always seem to preserve weak pullbacks. Note that while *pow* does preserve weak pullbacks it does not preserve pullbacks. It would be interesting to sort out the relationship between the two notions "uniform on maps" and "preserving weak pullbacks".

8 | An Application to Communicating Systems

In this chapter we illustrate some of the general theory described in the previous two chapters by considering the example from computer science of Robin Milner's Synchronous Calculus of Communicating Systems, abbreviated *SCCS*. (See Milner 1983.) This calculus can be viewed as a mathematically streamlined and synchronous version of the earlier calculus *CCS*. (See Milner 1980.) In (Milner 1983) Milner set up *SCCS* by giving an inductive definition of a class of infinitary expressions. These expressions are intended to represent the possible states of systems that can communicate with each other. Communication between systems is represented by synchronisation of atomic actions. To capture the idea of synchronisation Milner uses an Abelian group *Act* of atomic actions. The parallel synchronous composition of two atomic actions a, b is represented by the atomic action ab obtained by using the group operation to compose a and b. The identity $aa^{-1} = 1$, where 1 is the unit of the group, is used to represent the synchronisation of an atomic action a in one system with the inverse atomic action a^{-1} in another system. Here we take the view that this aspect of *SCCS* is not fundamental to its mathematics. So we will assume given an arbitrary set *Act* of atomic actions and impose no structure on it. Of course in the applications *Act* will need to be structured suitably, but such structure can be introduced as needed.

Milner gives the expressions of *SCCS* an operational semantics in terms of an inductive definition of a family of binary relations on the class of expressions. These relations are indexed by the set *Act* and used to represent allowed transition steps from one state of a system to another, each step being labelled with an element of *Act*. So the operational semantics determines what has been called a labelled transition system. Different expressions of *SCCS* can have the same abstract behaviour as determined by the operational semantics. In order to capture this notion of abstract behaviour Milner makes use of a concept first considered by David Park in (Park 1981). This is the concept of a bisimulation relation on a labelled transition system. Among the bisimulation relations there is always a maximal one, which is moreover an equivalence relation. The notion of bisimulation relation on a system used in this book is simply the special case of Park's notion when *Act* is a singleton set. Milner calls the maximal bisimulation relation on the expressions of *SCCS* *strong congruence*. The final step of Milner's construction is to form a quotient of the class of expressions by strong congruence. The result is a labelled transition system which gives a model of abstract behaviours for a certain notion of computational system.

As we will see, transition systems labelled by elements of a set *Act* can be viewed as coalgebras relative to the standard functor $pow(Act \times \cdots)$ and Milner's quotient construction then becomes a construction of a final coalgebra relative to that functor. In fact Milner's quotient construction was the prototype for a proof of the final coalgebra theorem. As final coalgebras are unique up to isomorphism when they exist, only the existence of a final coalgebra is of any purely mathematical concern. In fact the final coalgebra theorem applies to the functor, as it preserves weak pullbacks.

When *Act* is a singleton set then the functor is isomorphic to the functor *pow* whose final coalgebras are the complete systems used to model *AFA*. It was the initial perception of this connection between Milner's construction and set theory that has led to the author's interest in non-well-founded sets and the work presented in this book.

As just described, the connection between *SCCS* and *AFA* is that the model construction for *AFA* is simply a special case of the quotient construction for *SCCS*. Put another way, sets in the *AFA*-universe are the abstract behaviours for the special case of *SCCS* where there is only one atomic action.

In fact it is more profitable to reverse the connection between *SCCS* and *AFA* by making use of the special final coalgebra theorem. That theorem applies to the functor $pow(Act \times \cdots)$, as this standard functor is easily checked to be uniform on maps.

It follows that the largest fixed point of the functor is a final coalgebra, provided that we assume AFA. In this way we get a very simple and direct set theoretical construction which can be used to replace Milner's considerably more elaborate quotient construction. Of course the "penalty" to be paid for this simplicity is the need to use non-well-founded sets and AFA. But if one accepts the point of view suggested in this book then that is no penalty at all.

Transition Systems

Transition systems form a natural model for computation processes. Such systems consist of a class X of possible states of the system and a family of binary transition relations $\xrightarrow{a}$ between states, one for each possible atomic action a of a process. So $x \xrightarrow{a} y$ holds if there is a possible atomic transition step of the process from the state x to the state y in which the atomic action a takes place. So a computation of a process starting in a state x_0 will have the form $x_0 \xrightarrow{a_0} x_1 \xrightarrow{a_1} x_2 \xrightarrow{a_2} \cdots$ where $x_0, x_1, x_2, \ldots$ are the successive states that the process passes through and a_0, $a_1, a_2, \ldots$ are the atomic actions performed at successive steps of the computation. These atomic actions are intended to represent what is externally observable, while the successive states that a process passes through in a computation are intended to be internal to the process. So distinct states of processes may have the same external behaviour.

Transition systems may be conveniently represented as pairs (X, α) where X is a class and $\alpha : X \longrightarrow pow(Act \times X)$ is the map given by:

$$\alpha x \;=\; \{(a, y) \in Act \times X \mid x \xrightarrow{a} y\}$$

for all $x \in X$. The transition relations can be recaptured from α using the definitions:

$$x \xrightarrow{a} y \quad \Longleftrightarrow \quad (a, y) \in \alpha x$$

for $x, y \in X$. Now observe that $\alpha : X \longrightarrow \Theta X$ where Θ is the following standard functor on the category of classes:

$$\Theta \;=\; pow \circ (K_{Act} \times Id).$$

So from now on we will take a TS; i.e. a TRANSITION SYSTEM, to be a coalgebra relative to this functor Θ, with transition relations as determined above. Notice that we allow a *TS* to have a class of states and so will call it small if the class is in fact a set. We will call a coalgebra homomorphism between *TS*s a *TS* map.

The Complete Transition System $\mathcal{P}$

As Θ is a standard functor that preserves weak pullbacks we may apply the final coalgebra theorem to get the existence of a final coalgebra for Θ. We will call such a coalgebra a COM-PLETE *TS*. The abstract behaviours of *SCCS* turn out to be the states of a complete *TS*, a mathematical structure that is uniquely determined up to isomorphism. So *SCCS* could be developed axiomatically on the basis of a postulated complete *TS*. Here we prefer to follow an alternative course and instead use *AFA* and the special final coalgebra theorem. As the functor Θ is uniform on maps we can apply the theorem to get a simple set theoretical definition of a complete *TS* $\mathcal{P}$. $\mathcal{P}$ is defined to be the largest fixed point of Θ, or equivalently, it is the largest class such that if $P \in \mathcal{P}$ then P is a subset of $Act \times \mathcal{P}$. $\mathcal{P}$ is a *TS* with transition relations $\xrightarrow{a}$ for $a \in Act$ given by

$$P \xrightarrow{a} Q \quad \Longleftrightarrow \quad (a, Q) \in P$$

for all $P, Q \in \mathcal{P}$.

As $\mathcal{P}$ is a complete *TS*, for each *TS* (X, α) there is a unique *TS* map $(X, \alpha) \longrightarrow \mathcal{P}$. We will call this map the BEHAVIOUR MAP for (X, α). It is the unique map $\pi : X \longrightarrow \mathcal{P}$ such that for all $x \in X$

$$\pi x = \{(a, \pi y) \mid x \xrightarrow{a} y \text{ in } (X, \alpha)\}.$$

If a *TS* arises as an operational semantics for a programming language then the behaviour map for the *TS* will give a canonical representation of the abstract behaviours of the programs of the language, as given by the operational semantics. In this way the complete *TS* $\mathcal{P}$ plays the role of a domain of mathematical objects that can be the denotations of programs for such a programming language.

Some Operations on $\mathcal{P}$

We will define some operations on $\mathcal{P}$ that correspond to the four fundamental combinators that Milner used in (Milner 1983) to define the expressions of *SCCS*. The four combinators were called Action, Summation, Restriction and Product.

Action

We start with the action operations. Given $a \in Act$ there is an operation on $\mathcal{P}$ that assigns to each $P \in \mathcal{P}$ a set $a : P \in \mathcal{P}$ such that for all $b \in Act$ and $Q \in \mathcal{P}$

$$a : P \xrightarrow{b} Q \quad \Longleftrightarrow \quad [\, a = b \ \& \ P = Q \,].$$

So $a : P$ allows only the atomic action a to become P. In fact we define

$$a : P = \{(a, P)\}.$$

Summation

Next we consider the summation operations. Given $P_i \in \mathcal{P}$ for $i \in I$, where I is a set, there is a unique element $\sum_{i \in I} P_i$ of $\mathcal{P}$ such that for all $b \in Act$ and $Q \in \mathcal{P}$

$$\sum_{i \in I} P_i \xrightarrow{b} Q \quad \Longleftrightarrow \quad [\, P_i \xrightarrow{b} Q \text{ for some } i \in I \,].$$

In particular, when $I = \emptyset$ we get the null element $\mathbf{0}$ of $\mathcal{P}$ which allows no atomic steps, and when $I = \{1, \ldots, n\}$ we get the finite sum $P_1 + \ldots + P_n$. In fact in general we define

$$\sum_{i \in I} P_i \ = \ \bigcup_{i \in I} P_i$$

and in particular we get that

$$\mathbf{0} \ = \ \emptyset$$
$$P_1 + \cdots + P_n \ = \ P_1 \cup \cdots \cup P_n.$$

Note that the following equations trivially follow from these definitions.

$$P + Q = Q + P$$
$$P + (Q + R) = (P + Q) + R$$
$$P + \mathbf{0} = P$$
$$P + P = P$$

There are also equations for indexed sums that we do not bother to state as they are simply the expected equations for indexed unions of sets. See (Milner 1983) where such equations play an important role in understanding $SCCS$ as a calculus.

Restriction

The third operation on $\mathcal{P}$ that we define is the restriction operation. Given $A \subseteq Act$ there is a unique operation $- \restriction A : \mathcal{P} \longrightarrow \mathcal{P}$ such that for all $P \in \mathcal{P}$ if $b \in Act$ and $Q \in \mathcal{P}$ then $P \restriction A \xrightarrow{b} Q$ if and only if $b \in A$ and $P \xrightarrow{b} P'$ for some $P' \in \mathcal{P}$ such that $P' \restriction A = Q$. To see this we consider the TS $(\mathcal{P}, \alpha_A)$ where $\alpha_A : \mathcal{P} \longrightarrow pow(Act \times \mathcal{P})$ is given by

$$\alpha_A P \;=\; P \cap (A \times \mathcal{P})$$

for all $P \in \mathcal{P}$. Then we can define $- \restriction A : \mathcal{P} \longrightarrow \mathcal{P}$ to be the unique behaviour map for $(\mathcal{P}, \alpha_A)$.

Product

The product operation $\times$ is dependent on a binary composition operation assigning $ab \in Act$ to $a, b \in Act$. It is the unique binary operation on $\mathcal{P}$ such that for $P_1, P_2 \in \mathcal{P}$ if $b \in Act$ and $Q \in \mathcal{P}$ then $P_1 \times P_2 \xrightarrow{b} Q$ if and only if

$$P_1 \xrightarrow{a_1} Q_1 \quad \& \quad P_2 \xrightarrow{a_2} Q_2$$

for some $a_1, a_2 \in Act$ such that $a_1 a_2 = b$ and some $Q_1, Q_2 \in \mathcal{P}$ such that $Q_1 \times Q_2 = Q$. In fact we can define $\times$ to be the unique behaviour map for the TS $(\mathcal{P} \times \mathcal{P}, \gamma)$ where $\gamma : \mathcal{P} \times \mathcal{P} \longrightarrow pow(Act \times (\mathcal{P} \times \mathcal{P}))$ is given by

$$\gamma(P_1, P_2) \;=\; \{(a_1 a_2, (Q_1, Q_2)) \mid P_1 \xrightarrow{a_1} Q_1 \ \& \ P_2 \xrightarrow{a_2} Q_2\}$$

for all $P_1, P_2 \in \mathcal{P}$.

This completes the description of the operations on $\mathcal{P}$ corresponding to the four fundamental $SCCS$ combinators in (Milner 1983). Milner also considers two further derived combinators, morphism and delay. We give the operations on $\mathcal{P}$ that correspond to them.

Morphism

For each map $\varphi : Act \longrightarrow Act$ there is a unique morphism operation $-[\varphi] : \mathcal{P} \longrightarrow \mathcal{P}$ such that for all $P \in \mathcal{P}$ if $b \in Act$ and $Q \in \mathcal{P}$ then $P[\varphi] \xrightarrow{b} Q$ if and only if $P \xrightarrow{a} P'$ for some $a \in Act$ such that $\varphi a = b$ and some $P' \in \mathcal{P}$ such that $P'[\varphi] = Q$. In fact it is the unique behaviour map for the TS $(\mathcal{P}, \beta_\varphi)$ where $\beta_\varphi : \mathcal{P} \longrightarrow pow(Act \times \mathcal{P})$ is given by

$$\beta_\varphi P \;=\; \{(\varphi a, Q) \mid P \xrightarrow{a} Q\}$$

for all $P \in \mathcal{P}$.

Delay

The delay operation depends on a distinguished element $1 \in Act$. It is the unique operation $\delta : \mathcal{P} \longrightarrow \mathcal{P}$ such that for all $b \in Act$ and $Q \in \mathcal{P}$

$$\delta P \xrightarrow{b} Q \quad\Longleftrightarrow\quad [\,b = 1 \ \& \ \delta P = Q\,]\ \text{or}\ [\,P \xrightarrow{b} Q\,].$$

In fact we can define it to be the unique behaviour map for the TS $(\mathcal{P}, \sigma)$ where $\sigma : \mathcal{P} \longrightarrow pow(Act \times \mathcal{P})$ is given by

$$\sigma P \;=\; P \cup \{(1, P)\}$$

for all $P \in \mathcal{P}$.

As mentioned earlier the set Act is given the structure of an Abelian group in (Milner 1983). It is the group composition that is used in defining the product operation on $\mathcal{P}$. Also the unit 1 of the group is used in defining the delay operation. The restriction operation $- \upharpoonright A$ is only used when $1 \in A$ and the morphism operation $-[\varphi]$ is only used when $\varphi : Act \longrightarrow Act$ is a monoid homomorphism. The associative and commutative laws for the group operation on Act give rise to the same laws for the product operation; i.e.

$$P \times Q = Q \times P$$
$$P \times (Q \times R) = (P \times Q) \times R$$

for all $P, Q, R \in \mathcal{P}$. These laws are easily proved by making use of the uniqueness of behaviour maps. For example the associative law for $\times$ can be shown as follows. Let $\pi_1, \pi_2 : \mathcal{P} \times \mathcal{P} \times \mathcal{P} \longrightarrow \mathcal{P}$ be given by

$$\pi_1(P, Q, R) = P \times (Q \times R)$$
$$\pi_2(P, Q, R) = (P \times Q) \times R$$

for all $P, Q, R \in \mathcal{P}$. Now observe that both π_1 and π_2 are behaviour maps for the TS $(\mathcal{P} \times \mathcal{P} \times \mathcal{P}, \gamma')$ where $\gamma' : \mathcal{P} \times \mathcal{P} \times \mathcal{P} \longrightarrow pow(Act \times (\mathcal{P} \times \mathcal{P} \times \mathcal{P}))$ is given by

$$\gamma'(P, Q, R) = \{(abc, (P', Q', R')) \mid P \xrightarrow{a} P' \ \& \ Q \xrightarrow{b} Q' \ \& \ R \xrightarrow{c} R'\}$$

for all $P, Q, R \in \mathcal{P}$. Note that we have implicitly used the associativity of the group operation by leaving out brackets from the expression "abc". By the uniqueness of behaviour maps $\pi_1 = \pi_2$.

If we define $\mathbf{1} = \delta\mathbf{0}$ then it is the unique element of $\mathcal{P}$ such that

$$\mathbf{1} = \mathbf{1} : \mathbf{1}.$$

Also we have the equality

$$P \times \mathbf{1} = P$$

for all $P \in \mathcal{P}$. Note also the following distributivity laws where I is a set and $P_i \in \mathcal{P}$ for each $i \in I$.

$$Q \times (\sum_{i \in I} P_i) = \sum_{i \in I} (Q \times P_i)$$

$$(\sum_{i \in I} P_i) \restriction A = \sum_{i \in I} (P_i \restriction A)$$

$$(\sum_{i \in I} P_i)[\varphi] = \sum_{i \in I} P_i[\varphi]$$

There are a variety of other equations for these $SCCS$ operations which can be found in (Milner 1983).

Merge

Other operations on $\mathcal{P}$ can be defined as wanted by using variations on the definitions. For example we may wish to consider a parallel merge operation on $\mathcal{P}$ instead of the synchronous product $\times$ that has been defined. So let $- \mid - : \mathcal{P} \times \mathcal{P} \longrightarrow \mathcal{P}$ be the unique operation such that for all $P_1, P_2 \in \mathcal{P}$ if $b \in Act$ and $Q \in \mathcal{P}$ then $P_1 \mid P_2 \xrightarrow{b} Q$ if and only if either

$$P_1 \xrightarrow{b} P_1' \quad \text{and} \quad P_1' \mid P_2 = Q \text{ for some } P_1'$$

or else

$$P_2 \xrightarrow{b} P_2' \quad \text{and} \quad P_1 \mid P_2' = Q \text{ for some } P_2'.$$

It is the unique behaviour map for the TS $(\mathcal{P} \times \mathcal{P}, \mu)$ where for $P_1, P_2 \in \mathcal{P}$

$$\mu(P_1, P_2) = \{(a, (P_1', P_2)) \mid P_1 \xrightarrow{a} P_1'\} \cup \{(a, (P_1, P_2')) \mid P_2 \xrightarrow{a} P_2'\}.$$

Now each atomic step of $P_1 \mid P_2$ corresponds to an atomic step of exactly one of the processes P_1, P_2, with the other process not moving. This definition can be modified so as to allow for the synchronisation of an atomic action a of one process with the inverse atomic action a^{-1} of the other process. This can be done by replacing μ in the definition by the map μ' where, for all $P_1, P_2 \in \mathcal{P}$, $\mu'(P_1, P_2)$ is the union of the set $\mu(P_1, P_2)$ with the set

$$\{ (1, (P_1', P_2')) \mid [P_1 \xrightarrow{a} P_1' \ \& \ P_2 \xrightarrow{a^{-1}} P_2'] \text{ for some } a \in Act \}.$$

Appendices

A | Notes Towards a History

As indicated by the title this section is not intended to be a complete and scholarly historical review. When I first came to be interested in non-well-founded sets, not long ago, I knew very little about what had previously been written on the subject. Gradually, I became aware of the sporadic interest the idea had aroused in a variety of mathematicians throughout this century. It seemed worthwhile to attempt to give a review of the literature that I have become aware of, if only as a possible starting point for a future historian. I hope that what follows may also interest the more casual reader.

I find that the historical development of the idea of a non-well-founded set during this century can be conveniently divided into quarter century periods.

1900–1924 Development of the notion of a non-well-founded set.

1925–1949 The first order axiom of foundation, its relative consistency and independence.

1950–1974 Models of set theory without the axiom of foundation.

1975– Non-well-founded sets come of age.

Of course the above descriptions for each period only give a rough indication of some of what was going on.

1900–1924

From today's perspective it seems surprising that it took so long before mathematicians familiar with set theory developed an interest in the structure of the membership relation. It seems that it was only the jolt of Russell's paradox that initiated such an interest. For Cantor, even the idea of membership as a binary relation on a domain of objects seems to have been distant from his thinking. Consider Cantor's 1895 statement about his concept of set.*

> By a 'set' we understand every collection to a
> whole M of definite, well-differentiated objects m
> of our intuition or our thought.
> (We call these objects the 'elements' of M)
> (Cantor 1895, page 282)

It is not altogether clear from this statement alone that sets are themselves definite, well-differentiated objects and hence can themselves be elements. But there would seem to be little doubt that Cantor would have agreed that they were, if he had been asked. Nevertheless Cantor appears to have made little use of sets that have sets as elements. This is blatantly not the case for Frege and Russell who based their theory of the natural and transfinite numbers on equivalence classes of sets. For them natural numbers were sets of finite equinumerous sets.

Frege must have been the first to explicitly envisage a universe of objects, (for him *the* universe of all objects), including sets (for him the courses-of-values of propositional functions) with a binary membership relation on this universe. But he appears to have paid little attention to the structure of this membership relation. No doubt he was busy with more pressing tasks in completing his two volume work (Frege 1893). As it is, because of his combination of the course-of-values construction with his treatment of sentences as names of truth values his conception turned out to be incoherent, as demonstrated by Russell's paradox.

While Frege's approach to the notion of sets had received little attention from mathematicians, who were generally concerned with sets of objects of some specific kind e.g. sets of points, Russell's paradox must have drawn their attention to the possibility that sets could themselves be elements so that the question of the possible self membership of sets arises.

A variety of "solutions" to Russell's paradox were suggested, several by Russell himself. But Russell's preferred resolution was to use his theory of types. In that theory each object is always of

---

* I use the translation on page 33 of (Hallett 1984).

some unique type and sets of objects of a given type will themselves be objects of a distinct type. So while the theory does allow for a membership relation between objects of any given type and sets of such objects, it does not allow even for the meaningfulness of the assertion of the membership of a set in itself, as that would require the set to have distinct types.

Russell's theory of types had its own difficulties for mathematicians following the Cantorian tradition. Having once grasped the possibility from the presentation of Russell's paradox of having a domain of objects with a membership relation as framework for set theory, it was not long before an axiomatic approach to such a framework would be taken. And in (Zermelo 1908), the mainstream axiomatic approach to set theory was initiated. Naturally the debate continued among some mathematicians on whether it was possible for a set to be a member of itself. There is, for example, the debate between Eklund and Broden in the period 1915–1918, (see e.g. Eklund 1918*).

It was Mirimanoff, in (Miramanoff 1917a, 1917b), who formulated the fundamental distinction between the well-founded and non-well-founded sets. He called the well-founded sets 'ensembles ordinaire' and described the hierarchical structure of the universe of ordinary sets arranged according to their ordinal rank. He seems to have little qualms in accepting the existence of extraordinary sets such as those that have themselves as members. He even formulated a notion of isomorphism between possibly non-well-founded sets. When applied to the universe of pure sets this is the *tree-isomorphism* relation that holds between sets when their canonical tree pictures are isomorphic as trees. In spite of his significant contributions Mirimanoff does not formulate any axiom of foundation or anti-foundation or suggest any strengthening of the extensionality principle for sets using his tree isomorphism relation. Nevertheless one may already discern the beginnings of a realisation of the conceptual advantages to be gained by restricting attention to the universe of well-founded sets.

Zermelo's 1908 axiom system for set theory can be viewed as having its natural place among the axiomatisations of fundamental domains of mathematics such as the axioms for Euclidean Geometry, the Dedekind-Peano axioms for the natural numbers and

* I am grateful to M. Boffa for drawing my attention to these references.

the axioms for the completely ordered field of real numbers. In each of these earlier examples a certain 'extremal axiom' ensures that the axiom system is categorical, i.e. has a unique model, up to isomorphism. (Of course I am not concerned here with the modern idea of first order fully formalised axiomatisation, but rather the traditional informal idea).

Thus, in the case of the axiom system for the natural numbers, the extremal axiom is the principle of mathematical induction, which is a minimalisation axiom, as it expresses that no objects can be subtracted from the domain of natural numbers while keeping the truth of the other axioms. The axiom systems for Euclidean Geometry and the real numbers involve on the other hand completeness axioms. These are maximalisation axioms; i.e. they express that the domain of objects cannot be enlarged while preserving the truth of the other axioms.

In a natural move to 'complete' the axioms for set theory, so as to obtain a categorical axiomatisation, (Fraenkel 1922), introduces the idea of an axiom of restriction. This was to be a minimalisation axiom. Such an axiom would ensure that only sets actually required in order to satisfy the axioms would be in the domain of sets.

In particular this would rule out Mirimanoff's extraordinary sets. But it would also rule out those ordinary sets that are simply never obtained by repeatedly forming sets using the operations required by the axioms, for example, because their rank in the cumulative hierarchy is too high.

There are a number of difficulties in carrying out Fraenkel's objectives to reach a categorical axiomatisation, as was already pointed out in (Skolem 1922), and further emphasised in (von Neumann 1925). For a good modern discussion of these difficulties I refer the reader to (Fraenkel et al. 1973, §6.4), where two possible axioms of restriction are formulated and their inadequacy discussed. For a general discussion of extremal axioms see (Carnap and Bachmann 1936).

1925–1949

At the start of this period, in (von Neumann 1925), we see the first explicit formulation of an axiom expressing that all sets are well-founded. This was simply the assertion, in von Neumann's language of axiomatic set theory, that there is no infinite descending ∈-chain. Von Neumann introduced his axiom as a precise formulation of an axiom of restriction in Fraenkel's 1922 sense, realising full well that its addition to his axiom system would not make the system categorical.

The foundation axiom, *FA*, in its modern *ZFC*-form appears in (Zermelo 1930). Independently, von Neumann in (von Neumann 1929) had also presented essentially the same axiom as a reformulation of his 1925 axiom of restriction.

The relative consistency of *FA* with the axioms of set theory is also due to von Neumann. The result must have been unsurprising, as the inner model of the well-founded sets had already been introduced informally by Mirimanoff in 1917. But the relative independence of *FA* is more difficult and proofs of it did not appear until the 1950s although Bernays had already announced the result in (Bernays 1941).

Although Fraenkel's idea of a minimalising extremal axiom for set theory failed to give rise to a categorical axiom system it led eventually to the formulation of *FA*. It is in (Finsler 1926) that we see a formulation of an axiom system for set theory using an extremal axiom of the dual character of a maximalising axiom. This also fails to be a categorical axiom system having similar difficulties to Fraenkel's extremal axiom. Finsler appears to have been unresponsive to the criticisms of his idea.*

Nevertheless, his 1926 axiom system does lead to the formulation of what I have called Finsler's *AFA*. It is suprising that it has taken over 50 years for this "success" to come about, whereas Fraenkel only had to wait a handful of years. It is worth recording here that Finsler's axiom system uses a notion of isomorphism of sets which is different to the one introduced by Mirimanoff. If he had used Mirimanoff's notion the resulting anti-foundation axiom would have been what I have called Scott's *AFA*.

1950–1974

It is not until this period that we see proofs of the relative independence of the axiom of foundation. Bernay's proof of the result, while announced in 1941, did not appear until (Bernays 1954). Specker in his Habilitationschrift of 1951 gave a different proof, which may be found in (Specker 1957). In that paper may also be found an application of non-well-founded sets. This application

---

* See Finsler, 1975, for the long delayed second part of his 1926 paper.

uses reflexive sets i.e. sets such that $x = \{x\}$ so as to simulate Urelemente and so translate the Fraenkel-Mostowski method for independence results in set theory to a setting with only pure sets, admittedly sets that are possibly non-well-founded. But the very point of this simulation means that only a very limited kind of non-well-founded set is actually used; i.e. while there may be infinite descending $\in$-sequences, they all eventually become constantly equal to a reflexive set. Note that it is essential for the applications that infinitely many reflexive sets are needed so that the context is indeed some way from *AFA*, *SAFA* or *FAFA* where there is exactly one such set.

The methods of model construction for the independence result invented by Bernays turned out to be a very flexible tool for creating a great variety of models of set theory in which the axiom of foundation fails. Over the years the method was exploited by several people. (See Rieger 1957, Hájek 1965, Boffa 1969b, Felgner 1969 and especially Felgner's book, Felgner 1971, which gives a survey.) The general method is encapsulated in Rieger's theorem, (Rieger 1957). This result also covered Specker's construction, but the result has mostly been applied to systems V_π obtained by choosing a suitable permutation π of V.

In the same year as the important publications of Specker and Rieger we find in (Kanger 1957)* an unexpected role for non-well-founded sets in a completeness theorem for a variant of the predicate calculus. We have briefly explained this at the end of chapter 2. In his book Kanger states the set theoretical axiom he uses in an interesting "net" terminology for graphs. This terminology views a graph as a *net* made of *cords* tied together with *knots*. The cords are the nodes of the graph, while an edge arises when cords are tied together in a knot. Kanger suggests that this terminology has heuristic value in that the intuition underlying the formation of a set of objects is represented in an obvious manner by the act of tying the cords representing these objects together with a knot.

Dana Scott's (1960) contains a formulation of the axiom I have called *SAFA* and a model construction for it. Sadly this paper has remained unpublished. It was presented at the 1960 Stanford Congress and contains many interesting speculative

* I am grateful to Dag Westerståhl for drawing my attention to Kanger's book.

remarks.* Scott was unaware of (Specker 1957) when he wrote his paper, and preferred to publish another paper when he discovered that Specker had already given a similar model construction. Nevertheless, after (Finsler 1926), Scott appears to have been the first person to consider a strengthening of the axiom of extensionality. This idea seems to have then lain dormant until the 1980s.

Scott's model construction is in fact closely related to Specker's but there is a subtle difference in the notion of tree that they use. In fact neither of them formulate their notions of tree in terms of graphs but rather in terms of what it will be convenient here to call tree-partial-orderings. Scott's tree-partial-orderings are partially ordered sets having a largest element such that the sets of nodes larger than any given node form a finite chain under the orderings. Any tree T in the sense of this book determines such a tree-partial-ordering $\geq$ of the nodes by defining $a \geq b$ if and only if there is a path $a \to \cdots \to b$ in T (possibly of length 0, when $a = b$). Moreover, every tree-partial-ordering in Scott's sense arises in this way. Specker's notion of tree partial ordering† is, in fact, more general than Scott's. For Specker a partially ordered set $(A, \geq)$ is a tree partial ordering if it has a largest element such that the set of nodes at or above any given node form a chain in which every element of the chain is either the least element of the chain or else has an immediate predecessor in the chain. So Specker does not insist on these chains being finite or even being co-well-ordered.

The class of Specker tree partial orderings that have only a trivial automorphism form the nodes of a system M, where each node $(A, \geq)$ of M has as children those restrictions

$$(A_a, \{(x, y) \in A_a \times A_a \mid x \geq y\})$$

of $(A, \geq)$, where $a \in A$ is an immediate predecessor of the largest element of A. Here, for each such $a \in A$

$$A_a = \{x \in A \mid a \geq x\}.$$

Specker's model is then the full system obtained from M by forming a quotient of M with respect to the equivalence relation of isomorphism between the partially ordered trees in M.

---

* I am grateful to Robin Milner for sending me a copy of this paper that had been previously unknown to me.

† Here Specker's ordering is reversed, so as to be in line with Scott's.

Specker's model has quite different properties to Scott's model. There is a unique reflexive set in Scott's model, but there is a proper class of them in Specker's model. To see this observe that each ordinal α determines the tree partial ordering $(\alpha, \geq_\alpha) \in M$, where

$$x \geq_\alpha y \iff x < y < \alpha.$$

If the ordinal α is infinite then the tree partial ordering has a unique child in M which is isomorphic to it, so that it determines a reflexive set in the model. Moreover distinct infinite ordinals determine non-isomorphic tree partial orderings and hence distinct reflexive sets in the model.

The decade starting in 1965 witnessed a flurry of papers on non-well-founded sets exploiting the model construction techniques initiated by Bernays and Specker. There is (Hájek 1965) and a series of papers by Boffa listed in the references, as well as (Felgner 1969). As far as I am aware none of this work considers any strengthening of the extensionality axiom. Perhaps the highpoint of this period is Boffa's formulation of his axiom of superuniversality. This is the axiom that is called $BAFA$ here. The proof in chapter 5 that this axiom has a full model that is unique up to isomorphism is different to the original proof given by Boffa.

For a useful account of some of the work on non-well-founded sets up to 1971 see the book (Felgner 1971).

1975–

Boffa's axiom of superuniversality gave the strongest possible existence axiom for non-well-founded sets compatible with ZFC^-. Recent years has seen an interest in combining such an existence axiom with a strengthening of the extensionality axiom. In particular von Rimscha's axiom of strong extensionality, Sext, is the axiom I have called $FAFA_2$. (See von Rimscha 1981b, 1981c, 1983b.) Von Rimscha considers a variety of universality axioms including his axiom $U1$, which is what I have called $FAFA_1$. His axioms $U1$ and $U4$ are called BA_1 and GA by me. In his series of papers on non-well-founded sets listed in the references von Rimscha explores a variety of other interesting topics that have not been taken up here.

Von Rimscha's axiom Sext is based on the formulation of a notion of isomorphism between sets. As we have seen in this

book, the axiom of extensionality can be strengthened further
by using the maximal bisimulation relation between sets. The
notion of a maximal bisimulation relation and its use in constructing extensional models has been discovered independently
by many people. An early use may be found in (Friedman 1973)
in connection with versions of set theory that use intuitionistic
logic. This idea is carried further in (Gordeev 1982), where a
completeness axiom Cpl is formulated which we have called GA.
This axiom is a consequence of Boffa's axiom BA_1, but Gordeev
appears to have been unaware of Boffa's earlier work when he
wrote his paper. Hinnion also uses bisimulations to construct
extensional models, (see Hinnion 1980, 1981, 1986), but does not
formulate any axioms. Forti and Honsell formulate a number of
axioms and investigate their relationships in a series of papers
listed in the references. In particular their axiom X_1 in (Forti
and Honsell 1983) is the axiom I have called AFA.

I first came across maximal bisimulations in the work of Robin
Milner on mathematical models for concurrency. See Milner
(1980, 1983). These models involve labelled transition systems;
i.e. indexed families of binary relations, rather than the single
relation used in modelling the membership relation. The notion
of a bisimulation on a labelled transition system is due to David
Park (see Park 1981). In 1983, I was struck with the formal similarity between Milner's quotient construction for $SCCS$ and the
construction used by Friedman and then Gordeev. In seeking to
exploit this I was led to formulate the axiom AFA and then discovered in the summer of 1984 that the same axiom had already
been investigated by Forti and Honsell. As I got more interested
in non-well-founded sets I became aware of the earlier ideas and
sought to work out the relationships between them.

A natural way to try to understand non-well-founded sets
is to view them as limits, in some sense, of their well-founded
approximations. This approach is inspired by Scott's theory of
domains, but it cannot be done in any simple minded way, as
I found out. An approach to non-well-founded sets along these
lines was independently pursued by Lars Hallnäs and has led him
in (Hallnäs 1985) to a different looking construction of the essentially unique full model of AFA. His construction leads him to
consider an axiom that turns out to be equivalent to AFA. It appears that Hallnäs's construction has some advantages when seeking to model non-well-founded sets in a constructive framework.

In particular Ingrid Lindström, in (Lindström 1986), has worked out a version of the construction within Per Martin-Löf's Intuitionistic Theory of Types.

One of the most exciting areas of application for non-well-founded sets and *AFA* is situation semantics, a recent development of the model theoretical approach to the semantics of natural language.* Roughly, situations are taken to be parts of the world made up of facts. Each fact is made up of a relation, a tuple of objects appropriate for the relation and a polarity, and expresses that the tuple is in or is not in the relation depending on the polarity. As situations are themselves objects they can occur as components of facts; i.e. as objects in the tuple. So a situation can be a component of a fact that is in a situation and it is natural for circular situations to arise which contain facts about themselves. The straightforward way to give a set theoretical model for such a notion of situation is to represent situations as sets of facts and to represent a fact as a triple

$$(R, a, \sigma)$$

where R is the relation, a is the tuple appropriate for the relation and σ is the polarity 0 or 1. If this sort of set theoretical model is to be used then non-well-founded sets are essential if circular situations are to be represented. While the book (Barwise and Perry 1983) does not use non-well-founded sets they have been fully exploited in (Barwise and Etchemendy 1987). The latter book makes use of notions of circular situation and circular proposition to discuss the Liar paradox and uses non-well-founded sets to represent such abstract objects. Is *AFA* the appropriate axiom to use? In fact Barwise and Etchemendy use a version of $ZFC^- + AFA$ which allows for atoms. Could any of the other variants of *AFA* considered in this book be also exploited for similar purposes? This remains to be seen. Certainly *AFA* seems at present to be all that is needed for situation semantics.

---

* See (Barwise and Perry 1983) for the original book on the subject. See also (Barwise, 1985, 1986) for the relevance of non-well-founded sets.

B | Background Set Theory

Introduction

The aims of this appendix are to make clear to the reader how much knowledge of set theory is needed to understand this book, to catalogue the notation used that may not be standard and to present a proof of an important result due to Rieger that is not easily found elsewhere.

The reader will need to have seen something of the development of axiomatic set theory presented in textbooks such as (Enderton 1977, Halmos 1960). A summary of this material may be found in Chapter I of the excellent book (Kunen 1980). Chapters III and IV of that book form a convenient reference for additional material that it would be good for the reader to have seen. Certainly any reader who has read those chapters will find little difficulty with the contents of this book. Another worthwhile reference is (Shoenfield 1977).

I make free use of classes in this book, although I claim to be working informally in the axiomatic set theory, ZFC^-. The reader unfamiliar with this strategy should consult one of the above references. In part three familiarity with some of the language of category theory is needed. Very little standard category theory is really required, but the reader has to be prepared to consider functors on the *superlarge* category of classes. I found the book (Adámek 1983) helpful because it contains an investigation of certain types of functor on the category of sets.

Notation

The examples in chapter 1 of this book make use of the standard set theoretical representation of the natural numbers and ordered pairs. So the sets $\emptyset$, $\{\emptyset\}$, $\{\emptyset, \{\emptyset, \{\emptyset\}\}\}, \ldots$ are used to represent the natural numbers 0, 1, 2,... and in general the natural number n is represented by the set $\{m \mid m < n\}$ of natural numbers less than n. The ordered pair (a, b) is represented, as usual, by the set $\{\{a\}, \{a, b\}\}$, and the ordered n-tuple $(a_1, a_2, \ldots, a_{n-1}, a_n)$ can be represented, in terms of ordered pairs as $(a_1, (a_2, \ldots, (a_{n-1}, a_n)))$.

Many of the standard operations on sets carry over in a natural way to classes. So, for classes $A_1, \ldots A_n$ we have the classes

$$A_1 \cup \cdots \cup A_n,$$
$$A_1 \cap \cdots \cap A_n,$$
$$A_1 \times \cdots \times A_n$$

defined in the expected way. It will also be useful to have their disjoint union,

$$A_1 + \cdots + A_n = (\{1\} \times A_1) \cup \cdots \cup (\{n\} \times A_n).$$

For classes A, B their set difference will be written $A - B = \{x \in A \mid x \notin B\}$. The universal class of all sets is V. The power-class of a class A is the class $pow\, A = \{x \in V \mid x \subseteq A\}$ of all subsets of A.

A relation is a class of ordered pairs; i.e. R is a relation if $R \subseteq V \times V$. If R is a relation then xRy is written for $(x, y) \in R$ and the inverse of R is the relation $R^{-1} = \{(y, x) \mid xRy\}$. A relation R has domain $dom\, R = \{x \mid xRy \text{ for some } y\}$ and range $ran\, R = \{y \mid xRy \text{ for some } x\}$. The relational composition of relations R and S is the relation

$$R \mid S = \{(x, z) \mid xRy \ \& \ yRz \text{ for some } y\}.$$

The membership relation $\in$ is the class $\{(x, y) \mid x \in y\}$, and for each class A put $\in_A = \in \cap (A \times A)$.

For classes A, B a function $f : A \to B$ is a relation $f \subseteq A \times B$ such that for each $a \in A$ there is a unique $b \in B$ such that afb. This unique b is written fa or also $f(a)$. If $X \subseteq A$ then the restriction of f to X is the map $f \restriction X : X \to B$, given by $f \restriction X = f \cap (X \times B)$. If $Y \subseteq B$ then it's inverse image under f is $f^{-1}Y = \{x \in A \mid fx \in Y\}$. If $A \subseteq B$ and $f : A \to B$ such that $fx = x$ for all $x \in A$ then f is an inclusion map and is written $f : A \hookrightarrow B$. If $f : A \to B$ and $g : B \to C$ then their function composition $g \circ f : A \to C$ is given by

$$(g \circ f)x = g(fx) \qquad \text{for all } x \in A.$$

If A is a class and I is a set then A^I is the class of all the functions $f : I \to A$.

If A is a class of sets then

$$\bigcup A = \{x \mid x \in a \text{ for some } a \in A\},$$

$$\bigcap A = \{x \mid x \in a \text{ for all } a \in A\}.$$

For each class I a family of classes, A_i for $i \in I$, indexed by the class I can be represented as a relation $A \subseteq I \times V$, with $A_i = \{x \mid iAx\}$ for each $i \in I$. Given such a family of classes

$$\bigcup_{i \in I} A_i = \{x \mid x \in A_i \text{ for some } i \in I\},$$

$$\bigcap_{i \in I} A_i = \{x \mid x \in A_i \text{ for all } i \in I\},$$

$$\sum_{i \in I} A_i = \bigcup_{i \in I} \{i\} \times A_i,$$

and if I is a set,

$$\prod_{i \in I} A_i = \{f \in (\bigcup_{i \in I} A_i)^I \mid fi \in A_i \text{ for all } i \in I\}.$$

Occasionally it is convenient to consider mathematical structures having a proper class A as universe. It is usual to keep to the usual tupling notation $(A, \ldots)$ for such a structure, even though the standard definition of tuples only applies to sets. This can be understood as the class $A + R + \cdots$, using the disjoint union operation. A class that is actually a set is also called a small class, and a structure whose universe is small is called a small structure. All the functions and relations that make up a small structure will also be small.

In part III set continuous operators Φ are used. These assign a class ΦX to each class X. Because of the set continuity property the operator can be represented as the class

$$\hat{\Phi} = \{(a, x) \mid a \in \Phi x\},$$

as then for each class X

$$\Phi X = \{a \mid a\hat{\Phi}x \text{ for some } x \in pow X\}.$$

Well-Foundedness

A relation R is well-founded if there is no infinite sequence a_0, $a_1, \ldots$ such that $a_{n+1} R a_n$ for $n = 0, 1, \ldots$. A set a is well-founded if there is no infinite sequence a_0, $a_1, \ldots$ such that $a_0 \in a$ and $a_{n+1} \in a_n$ for $n = 0, 1, \ldots$. V_{wf} is the class of all the well-founded sets.

A class A is transitive if $A \subseteq pow A$; i.e. every element of A is a subset of A. For transitive classes A we have the following principles, provided that the elements of A are all well-founded sets.

Set Induction on A:
For any class B if

$$a \subseteq B \implies a \in B \text{ for all } a \in A$$

then $A \subseteq B$.

Set Recursion on A:
To uniquely define $F : A \to V$ it suffices to define Fa in terms of $F \upharpoonright a$ for each $a \in A$.

The following important result plays a special role in chapter 1.

Mostowsky's Collapsing Lemma:
If R is a well-founded relation on the set A then there is a unique function $f : A \to V$ such that for all $a \in A$

$$fa = \{fx \mid xRa\}.$$

The assumption that the class A is a set can be dropped provided it is assumed instead that $\{x \mid xRa\}$ is a set for each $a \in A$; i.e. that (A, R) is a system in the sense of chapter 1.

We will use the standard von Neumann treatment of the ordinals, so that an ordinal α is identified with the set $\{\beta \mid \beta < \alpha\}$ of it's predecessors. So the class On of ordinals is defined to be the class of well-founded transitive sets, all of whose elements are also transitive.

The Axiomatisation of Set Theory

We take a standard first order language for set theory that just has the binary predicate symbols '=' and '∈'. We assume a standard axiomatisation of first order logic with equality. Also we use the standard abbreviations for the restricted quantifiers

$$\forall x \in a \cdots \quad \overset{\text{def}}{=} \quad \forall x(x \in a \rightarrow \cdots),$$

$$\exists x \in a \cdots \quad \overset{\text{def}}{=} \quad \exists x(x \in a \ \& \ \cdots).$$

In the following list of non-logical axioms for ZFC^- we have avoided the use of any other abbreviations.

Extensionality:
$$\forall z(z \in a \leftrightarrow z \in b) \rightarrow a = b$$

Pairing:
$$\exists z[\, a \in z \ \& \ b \in z \,]$$

Union:
$$\exists z(\forall x \in a)(\forall y \in x)(y \in z)$$

Powerset:
$$\exists z \forall x[\, (\forall u \in x)(u \in a) \ \rightarrow \ x \in z \,]$$

Infinity:
$$\exists z[\, (\exists x \in z)\forall y \neg(y \in x) \ \& \ (\forall x \in z)(\exists y \in z)(x \in y) \,]$$

Separation:
$$\exists z \forall x[\, x \in z \ \leftrightarrow \ x \in a \ \& \ \varphi \,]$$

Collection:
$$(\forall x \in a)\exists y \, \varphi \ \rightarrow \ \exists z(\forall x \in a)(\exists y \in z) \, \varphi$$

Choice:
$$(\forall x \in a)\exists y(y \in x)$$
$$\& \ (\forall x_1 \in a)(\forall x_2 \in a)[\, \exists y(y \in x_1 \ \& \ y \in x_2) \rightarrow \ x_1 = x_2 \,]$$
$$\rightarrow \exists z(\forall x \in a)(\exists y \in x)(\forall u \in x)[\, u \in z \leftrightarrow u = y \,]$$

The choice axiom is abbreviated AC. Separation and Collection are schemes in which φ can be any formula in which the variable z does not occur free. ZFC is ZFC^- together with the following axiom.

Foundation:
$$\exists x (x \in a) \;\rightarrow\; (\exists x \in a)(\forall y \in x)\neg(y \in a).$$

This axiom is abbreviated FA.

ZFC has usually been formulated using the axiom scheme of replacement rather than the collection scheme used here. This makes no difference to the theorems of ZFC, but it probably does to the theorems of ZFC^-, as while each instance of replacement can easily be proved from collection, the usual proof of each instance of collection in ZFC makes essential use of FA. I prefer to take the apparently stronger collection scheme.

Global Choice and Quotients

When working with classes it is sometimes convenient to be able to use a global form of AC. The form that is used in this book is

$$V \cong On.$$

This expresses that there is a bijection between the universe and the class On of ordinals. This axiom cannot be formulated in the language of set theory alone but an additional predicate symbol is needed for the bijection and the axiom schemes of ZFC^- need to be extended to the larger language. A fairly cavalier approach to the use of AC is taken in this book. The stronger global form is used whenever it appears needed. One use of global choice is in the formation of the quotient of a class by an equivalence relation. In many situations this use can be avoided. If R is an equivalence relation on the class A we will call $f : A \to B$ a quotient of A with respect to R if f is surjective and for all $a_1, a_2 \in A$
$$a_1 R a_2 \;\Longleftrightarrow\; f a_1 = f a_2.$$

Using global AC a quotient can be obtained as follows. The bijection between V and On determines a well-ordering of V. For each $a \in A$ let fa be the least set b in the well-ordering such that $b \in A$ and aRb. When A is a set, or more generally when each equivalence class $\{x \mid xRa\}$ is a set, we can follow the familiar procedure of defing fa to be the equivalence class of a. This method works in ZF^-; i.e. ZFC without FA or AC. For equivalence relations on a class A in general there is a trick to get a quotient, due to Dana Scott, that makes essential use of FA. The trick is to define fa to be the subset of the equivalence class $\{x \mid xRa\}$ consisting of those elements of the equivalence class having the least possible rank in the cumulative hierarchy of well-founded sets. In ZFC^- this trick is no longer available, but often a slight variation of the trick will work. For example if A is the class of linearly ordered sets and R is the isomorphism relation between linearly ordered sets then if $a \in A$ we can let fa be the set of linear orderings of the ordinal α that are isomorphic to the linearly ordered set a, where α is the least possible ordinal for which there is such a linear ordering of α. This works because by AC every set is in one-one correspondence with an ordinal.

Rieger's Theorem

Here we will prove the result that gives a general method for giving interpretations of ZFC^-. In order to interprete the language of set theory all that is needed is a class M for the variables to range over and a binary relation $\in_M \subseteq M \times M$ to interprete the predicate symbol '$\in$'. Now any system M, in the sense of chapter 1, determines the binary relation $\in_M$ given by

$$a \in_M b \iff a \in b_M.$$

We will show that this gives an interpretation of all the axioms of ZFC^- provided that the system is full. Recall from chapter 3 that a system M is full if for each set $x \subseteq M$ there is a unique $a \in M$ such that $x = a_M$. In the following we will let x^M be this unique $a \in M$.

Rieger's Theorem:
Every full system is a model of ZFC^-.

Proof: Let M be a full system. We will consider each axiom of ZFC^- in turn.

- Extensionality: Let $a, b \in M$ such that

$$M \models \forall x (x \in a \leftrightarrow x \in b).$$

Then $a_M = b_M$, so that $a = (a_M)^M = (b_M)^M = b$ and hence $M \models a = b$.

- Pairing: If $a, b \in M$ then $c = \{a, b\}^M \in M$ is such that $M \models (a \in c \,\&\, b \in c)$.
- Union: Let $a \in M$. Then $\bigcup \{y_M \mid y \in a_M\}$ is a subset x of M so that if $c = x^M \in M$ then $M \models \forall y \in a \forall z \in y (z \in c)$.
- Powerset: If $a \in M$ then $c = \{x^M \mid x \subseteq a_M\}^M \in M$ is such that
$$M \models \forall x [\, \forall z \in x (z \in a) \rightarrow x \in c \,].$$

- Infinity: Let
$$\begin{cases} \Delta_0 = \emptyset^M \\ \Delta_{n+1} = ((\Delta_n)_M \cup \{\Delta_n\})^M \text{ for } n = 0, 1, \dots \end{cases}$$

Then $\Delta_n \in M$ for each natural number n, so that
$$\Delta_\omega = \{\Delta_n \mid n = 0, 1, \dots\}^M \in M$$

is such that
$$M \models [\, \Delta_0 \in \Delta_\omega \,\&\, \forall y (y \notin \Delta_0) \,]$$

and
$$M \models \forall x \in \Delta_\omega \exists y \in \Delta_\omega (x \in y).$$

- Separation: Let $a \in M$ and let φ be a formula containing at most x free and perhaps constants for elements of M. Then
$$c = \{b \in a_M \mid M \models \varphi[b/x]\}^M \in M$$

is such that
$$M \models \forall x (x \in c \,\leftrightarrow\, x \in a \,\&\, \varphi).$$

- Collection: Let $a \in M$ and let φ be a formula containig at most x and y free and perhaps constants for elements of M. Suppose that
$$M \models \forall x \in a \exists y \, \varphi.$$
Then
$$\forall x \in a_M \exists y [\, y \in M \,\&\, M \models \varphi \,].$$

By the collection axiom scheme there is a set b such that
$$\forall x \in a_M \exists y \in b [\, y \in M \,\&\, M \models \varphi \,].$$

As $b \cap M$ is a subset of M we may form $c = (b \cap M)^M \in M$ such that
$$M \models \forall x \in a \exists y \in c \, \varphi.$$

- Choice: Let $a \in M$ such that
$$M \models \forall x \in a \exists y (y \in x)$$

and

$$M \models (\forall x_1, x_2 \in a)[\, \exists y (y \in x_1 \ \& \ y \in x_2) \ \rightarrow \ x_1 = x_2 \,].$$

Then
$$\forall x \in a_M \ x_M \neq \emptyset$$
and for all $x_1, x_2 \in a_M$

$$(x_1)_M \cap (x_2)_M \neq \emptyset \implies x_1 = x_2.$$

Thus $\{x_M \mid x \in a_M\}$ is a set of non-empty pairwise disjoint sets. Hence by the axiom of choice there is a set b such that for each $x \in a_M$ the set $b \cap x_M$ has a unique element $c_x \in M$. Hence $c = \{c_x \mid x \in a_M\} \in M$ such that

$$M \models \forall x \in a \exists y \in x \forall u \in x [\, u \in c \leftrightarrow u = y \,].$$

$\square$

References

Adámek, J. 1983. *Theory of Mathematical Structures*. Dordrecht/ Boston/Lancaster: D. Reidel.

Barwise, J. 1975. *Admissible Sets and Structures*. Berlin/Heidelberg/ New York: Springer-Verlag.

Barwise, J. 1985. Modeling Shared Understanding. CSLI Informal Notes, Stanford University.

Barwise, J. 1986. Situations, Sets and the Axiom of Foundation. In Paris, Wilkie, and Wilmers (Eds.), *Logic Colloquium '84*, North Holland. 21–36.

Barwise, J. and J. Etchemendy. 1987. *The Liar: An Essay on Truth and Circular Propositions*. Oxford: Oxford University Press.

Barwise, J. and J. Perry. 1983. *Situations and Attitudes*. Cambridge, Mass. and London: MIT Press.

Bernays, P. 1941. A System of Axiomatic Set Theory, II. *J. Symbolic Logic* 6:1–17.

Bernays, P. 1954. A System of Axiomatic Set Theory, VII. *J. Symbolic Logic* 19:81–96.

Boffa, M. 1967a. Modèles de la théorie des Ensembles, associés aux permutations de l'univers. *Contes Rendus Acad. Sciences, Paris, Serie A* 264:221–222.

Boffa, M. 1967b. Remarques concernant les modèles associés aux permutations de l'univers. *Contes Rendus Acad. Sciences, Paris, Serie A* 265:205–206.

Boffa, M. 1968a. Graphes extensionelles et axiome d'universitalité. *Zeitschrift für math. Logik und Grundlagen der Math.* 14:329–334.

Boffa, M. 1968b. Les ensembles extraordinaire. *Bull. Soc. Math. Belg.* 20:3–15.

Boffa, M. 1969a. Axiome et schema de fondement dans le systeme de Zermelo. *Bull. Acad. Polon. Scie.* 17:113–115.

Boffa, M. 1969b. Sur la théorie des ensembles sans axiome de Fondement. *Bull. Soc. Math. Belg.* 31:16–56.

Boffa, M. 1971. Forcing et reflection. *Bull. Acad. Polon. Sci., Ser. Sci. Math.* 19:181–183.

Boffa, M. 1972a. Forcing et negation de l'axiome de Fondement. *Memoire Acad. Sci. Belg.* tome XL, fasc. 7.

Boffa, M. 1972b. Formules Σ_1 en théorie des ensembles sans axiome de fondement. *Zeitschrift für math. Logik und Grundlagen der Math.* 18:93–96.

Boffa, M. 1973. Structures extensionelles Generiques. *Bull. Soc. Math. Belg.* 25:1–10.

Boffa, M. and J. Beni. 1968. Elimination de cycles d'appertenance par permutation de l'univers. *Contes Rendus Acad. Sciences, Paris, Serie A* 266:545–546.

Boffa, M. and G. Sabbagh. 1970. Sur l'axiome U de Felgner. *Contes Rendus Acad. Sciences, Paris, Serie A* 270:993–994.

Cantor, G. 1895. Beiträge zur Begründung der transfiniten Mengenlehre, 1. *Mathematische Annalen* 46:481–512.

Carnap, R. and F. Bachmann. 1936. Über Extremalaxiome. *Erkenntnis* 6:166–188. English Translation in *History and Philosophy of Logic*, 2:67–85. (1981).

Eklund, H. 1918. Über Mengen, die Elemente ihrer selbst sind. *Nyt Tidsskrift for Matematik* Afdeling B:8–28.

Enderton, H. B. 1977. *Elements of Set Theory*. New York: Academic Press.

Felgner, U. 1969. Die Inklusionsrelation zwischen Universa und ein abgeschwachtes Fundierungsaxiom. *Archiv. der Math. Logik* 20:561–566.

Felgner, U. 1971. *Models of ZF Set Theory*. Berlin: Springer. Springer Lecture Notes in Mathematics No. 223.

Finsler, P. 1926. Über die Grundlagen der Mengenlehre, I. *Math. Zeitschrift* 25:683–713. Appears in Finsler (1975). Also appearing there is part II, originally published in Comment. Math. Helv., Vol. 38, 1964, 172–218.

Finsler, P. 1975. *Aufsatse zur Mengenlehre.* Darmstadt: Wissenschaftliche Buchgesellschaft.

Forti, M. and F. Honsell. 1983. Set Theory with Free Construction Principles. *Annali Scuola Normale Superiore—Pisa Classe di Scienza* 10:493–522. Serie IV.

Forti, M. and F. Honsell. 1984a. Axioms of Choice and Free Construction Principles, I. *Bull. Soc. Math. Belg.* 36:69–79. Part II, 37:1–16. Part III, to appear.

Forti, M. and F. Honsell. 1984b. Comparison of the Axioms of Local and Global Universality. *Zeitschrift für math. Logik und Grundlagen der Math.* 30:193–196.

Forti, M. and F. Honsell. 1985a. The Consistency of the Axiom of Universality for the Ordering of Cardinals. *J. Symbolic Logic* 50:502–509.

Forti, M. and F. Honsell. 1985b. A Model Where Cardinal Ordering is Universal. *Zeitschrift für math. Logik und Grundlagen der Math.* 31:533–536.

Fraenkel, A. 1922. Zu den Grundlagen der Cantor-Zermeloschen Mengenlehre. *Mathematische Annalen* 86:230–237.

Fraenkel, A., Y. Bar-Hillel, and A. Lévy. 1973. *Foundations of Set Theory.* Amsterdam: North Holland.

Frege, G. 1893. *Grundgesetze der Arithmetik, begriffsschriftlich abgeleitet, Vol. 1.* Jena. Volume 2 published in 1903.

Friedman, H. 1973. The Consistency of Classical Set Theory Relative to a Set Theory with Intuitionistic Logic. *J. Symbolic Logic* 38:315–319.

Gordeev, L. 1982. Constructive Models for Set Theory with Extensionality. In A. S. Troelstra and D. van Dalen (Eds.), *The L.E.J. Brouwer Centenary Symposium,* North Holland. 123–147.

Hájek, P. 1965. Modelle der Mengenlehre in den Mengen gegebener Gestalt existieren. *Zeitschrift für math. Logik und Grundlagen der Math.* 11:103–115.

Hallett, M. 1984. *Cantorian Set Theory and Limitation of Size.* Oxford: Clarendon Press.

Hallnäs, L. 1985. Approximations and Descriptions of Non-well-founded Sets. Preprint, Department of Philosophy, University of Stockholm.

Halmos, P. R. 1960. *Naive Set Theory.* New York: Van Nostrand Reinhold.

Hinnion, R. 1980. Contraction de structures et application a NFU. *Commptes Rendus de l'Acad. des Sciences de Paris* 290:677–680.

Hinnion, R. 1981. Extensional Quotients of Structures and Applications to the Study of the Axiom of Extensionality. *Bull. de la Soc. Math. de Belg.* 33:173–206. Fasc. II, Ser. B.

Hinnion, R. 1986. Extensionality in Zermelo-Fraenkel Set Theory. *Zeitschrift für math. Logik und Grundlagen der Math.* 32:51–60.

Kanger, S. 1957. *Provability in Logic.* University of Stockholm: Almqvist and Wiksell. Stockholm Studies in Philosophy 1.

Kunen, K. 1980. *Set Theory.* Amsterdam: North Holland.

Lindström, I. 1986. A Construction of Non-well-founded Sets Within Martin-Löf's Type Theory. Report No. 15, Department of Mathematics, Uppsala University.

Milner, R. 1980. *A Calculus of Communicating Systems.* Berlin: Springer-Verlag. Lecture Notes in Computer Science, No. 92.

Milner, R. 1983. Calculi for Synchrony and Asynchrony. *Theoretical Computer Science* 25:267–310.

Mirimanoff, D. 1917a. Les antinomies de Russell et de Burali-Forti et le probleme fondamental de la theorie des ensembles. *L'enseignment mathematique* 19:37–52.

Mirimanoff, D. 1917b. Remarques sur la theorie des ensembles. *L'enseignment mathematique* 19:209–217.

Park, D. 1981. Concurrency and Automata on Infinite Sequences. In *Proceedings of the 5th GI Conference*, Springer. Springer Lecture Notes in Computer Science, No. 104. 167–183.

Rieger, L. 1957. A Contribution to Gödel's Axiomatic Set Theory, I. *Czechoslovak Math. J.* 7:323–357.

Scott, D. 1960. A Different Kind of Model for Set Theory. Unpublished paper given at the 1960 Stanford Congress of Logic, Methodology and Philosophy of Science.

Shoenfield, J. R. 1977. Axioms of Set Theory. In J. Barwise (Ed.), *Handbook of Mathematical Logic.* Amsterdam: North Holland. 321–344.

Skolem, T. 1922. Einige Bemerkungen zur axiomatischen Begründung der Mengenlehre. English translation in van Heijenoort (1967).

Specker, E. 1957. Zur Axiomatic der Mengenlehre (Fundierungsaxiom und Auswahlaxiom). *Zeitschrift für math. Logik und Grundlagen der Math.* 3:173–210.

van Heijenoort, J. (Ed.). 1967. *From Frege to Gödel*. Cambridge, Mass.: Harvard University Press.

von Neumann, J. 1925. Eine Axiomatisierung der Mengenlehre. *Journal für die reine und angewandte Mathematik* 154:219–240. English translation in van Heijenoort (1967) pages 393–413.

von Neumann, J. 1929. Über eine Widerspruchfreiheitsfrage in der axiomatischen Mengenlehre. *Journal für die reine und angewandte Mathematik* 160:227–241.

von Rimscha, M. 1978. *Nichtfundierte Mengenlehre*. PhD thesis, Kiel.

von Rimscha, M. 1980. Mengentheoretische Modelle des λK-Kalkuls. *Archiv für math. Logik und Grundlagenforschung* 20:65–73.

von Rimscha, M. 1981a. Das Kollektionsaxiom. *Zeitschrift für math. Logik und Grundlagen der Math.* 27:189–192.

von Rimscha, M. 1981b. Universality and Strong Extensionality. *Archiv für math. Logik und Grundlagenforschung* 21:179–193.

von Rimscha, M. 1981c. Weak Foundation and Axioms of Universality. *Archiv für math. Logik und Grundlagenforschung* 21:195–205.

von Rimscha, M. 1982. Transitivitatsbedingungen. *Zeitschrift für math. Logik und Grundlagen der Math.* 28:67–74.

von Rimscha, M. 1983a. Bases of ZF^0-models. *Archiv für math. Logik und Grundlagenforschung* 23:11–19.

von Rimscha, M. 1983b. Hierarchies for Non-well-founded Models of Set Theory. *Zeitschrift für math. Logik und Grundlagen der Math.* 29:253–288.

Zermelo, E. 1908. Untersuchungen über die Grundlagen der Mengenlehre, I. *Mathematische Annalen* 65:261–281. English Translation in van Heijenoort (1967).

Zermelo, E. 1930. Über Grenzzahlen und Mengenbereiche. *Fundamenta mathematicae* 16:29–47.

Index of Definitions

Index of Named Axioms and Results

CSLI Publications

Reports

The following titles have been published in the CSLI Reports series. These reports may be obtained from CSLI Publications, Ventura Hall, Stanford University, Stanford, CA 94305-4115.

The Situation in Logic–I Jon Barwise CSLI–84–2 (*$2.00*)

Coordination and How to Distinguish Categories Ivan Sag, Gerald Gazdar, Thomas Wasow, and Steven Weisler CSLI–84–3 (*$3.50*)

Belief and Incompleteness Kurt Konolige CSLI–84–4 (*$4.50*)

Equality, Types, Modules and Generics for Logic Programming Joseph Goguen and José Meseguer CSLI–84–5 (*$2.50*)

Lessons from Bolzano Johan van Benthem CSLI–84–6 (*$1.50*)

Self-propagating Search: A Unified Theory of Memory Pentti Kanerva CSLI–84–7 (*$9.00*)

Reflection and Semantics in LISP Brian Cantwell Smith CSLI–84–8 (*$2.50*)

The Implementation of Procedurally Reflective Languages Jim des Rivières and Brian Cantwell Smith CSLI–84–9 (*$3.00*)

Parameterized Programming Joseph Goguen CSLI–84–10 (*$3.50*)

Morphological Constraints on Scandinavian Tone Accent Meg Withgott and Per-Kristian Halvorsen CSLI–84–11 (*$2.50*)

Partiality and Nonmonotonicity in Classical Logic Johan van Benthem CSLI–84–12 (*$2.00*)

Shifting Situations and Shaken Attitudes Jon Barwise and John Perry CSLI–84–13 (*$4.50*)

Aspectual Classes in Situation Semantics Robin Cooper CSLI–85–14-C (*$4.00*)

Completeness of Many-Sorted Equational Logic Joseph Goguen and José Meseguer CSLI–84–15 (*$2.50*)

Moving the Semantic Fulcrum Terry Winograd CSLI–84–17 (*$1.50*)

On the Mathematical Properties of Linguistic Theories C. Raymond Perrault CSLI–84–18 (*$3.00*)

A Simple and Efficient Implementation of Higher-order Functions in LISP Michael P. Georgeff and Stephen F.Bodnar CSLI–84–19 (*$4.50*)

On the Axiomatization of "if-then-else" Irène Guessarian and José Meseguer CSLI–85–20 (*$3.00*)

The Situation in Logic–II: Conditionals and Conditional Information Jon Barwise CSLI–84–21 (*$3.00*)

Principles of OBJ2 Kokichi Futatsugi, Joseph A. Goguen, Jean-Pierre Jouannaud, and José Meseguer CSLI–85–22 (*$2.00*)

Querying Logical Databases Moshe Vardi CSLI–85–23 (*$1.50*)

Computationally Relevant Properties of Natural Languages and Their Grammar Gerald Gazdar and Geoff Pullum CSLI–85–24 (*$3.50*)

An Internal Semantics for Modal Logic: Preliminary Report Ronald Fagin and Moshe Vardi CSLI–85–25 (*$2.00*)

The Situation in Logic–III: Situations, Sets and the Axiom of Foundation Jon Barwise CSLI–85–26 (*$2.50*)

Semantic Automata Johan van Benthem CSLI–85–27 (*$2.50*)

Restrictive and Non-Restrictive Modification Peter Sells CSLI–85–28 (*$3.00*)

Institutions: Abstract Model Theory for Computer Science J. A. Goguen and R. M. Burstall CSLI–85–30 (*$4.50*)

134

Tarski on Truth and Logical Consequence John Etchemendy CSLI–86–64 (*$3.50*)

The LFG Treatment of Discontinuity and the Double Infinitive Construction in Dutch Mark Johnson CSLI–86–65 (*$2.50*)

Categorial Unification Grammars Hans Uszkoreit CSLI–86–66 (*$2.50*)

Generalized Quantifiers and Plurals Godehard Link CSLI–86–67 (*$2.00*)

Radical Lexicalism Lauri Karttunen CSLI–86–68 (*$2.50*)

Understanding Computers and Cognition: Four Reviews and a Response Mark Stefik, Editor CSLI–87–70 (*$3.50*)

The Correspondence Continuum Brian Cantwell Smith CSLI–87–71 (*$4.00*)

The Role of Propositional Objects of Belief in Action David J. Israel CSLI–87–72 (*$2.50*)

From Worlds to Situations John Perry CSLI–87–73 (*$2.00*)

Two Replies Jon Barwise CSLI–87–74 (*$3.00*)

Semantics of Clocks Brian Cantwell Smith CSLI–87–75 (*$2.50*)

Varieties of Self-Reference Brian Cantwell Smith CSLI–87–76 (*Forthcoming*)

The Parts of Perception Alexander Pentland CSLI–87–77 (*$4.00*)

Topic, Pronoun, and Agreement in Chicheŵa Joan Bresnan and S. A. Mchombo CSLI–87–78 (*$5.00*)

HPSG: An Informal Synopsis Carl Pollard and Ivan A. Sag CSLI–87–79 (*$4.50*)

The Situated Processing of Situated Language Susan Stucky CSLI–87–80 (*Forthcoming*)

Muir: A Tool for Language Design Terry Winograd CSLI–87–81 (*$2.50*)

Final Algebras, Cosemicomputable Algebras, and Degrees of Unsolvability Lawrence S. Moss, José Meseguer, and Joseph A. Goguen CSLI–87–82 (*$3.00*)

The Synthesis of Digital Machines with Provable Epistemic Properties Stanley J. Rosenschein and Leslie Pack Kaelbling CSLI–87–83 (*$3.50*)

Formal Theories of Knowledge in AI and Robotics Stanley J. Rosenschein CSLI–87–84 (*$1.50*)

An Architecture for Intelligent Reactive Systems Leslie Pack Kaelbling CSLI–87–85 (*$2.00*)

Order-Sorted Unification José Meseguer, Joseph A. Goguen, and Gert Smolka CSLI–87–86 (*$2.50*)

Modular Algebraic Specification of Some Basic Geometrical Constructions Joseph A. Goguen CSLI–87–87 (*$2.50*)

Persistence, Intention and Commitment Phil Cohen and Hector Levesque CSLI–87–88 (*$3.50*)

Rational Interaction as the Basis for Communication Phil Cohen and Hector Levesque CSLI–87–89 (*Forthcoming*)

An Application of Default Logic to Speech Act Theory C. Raymond Perrault CSLI–87–90 (*$2.50*)

Models and Equality for Logical Programming Joseph A. Goguen and José Meseguer CSLI–87–91 (*$3.00*)

Order-Sorted Algebra Solves the Constructor-Selector, Mulitple Representation and Coercion Problems Joseph A. Goguen and José Meseguer CSLI–87–92 (*$2.00*)

Extensions and Foundations for Object-Oriented Programming Joseph A. Goguen and José Meseguer CSLI–87–93 (*$3.50*)

L3 Reference Manual: Version 2.19 William Poser CSLI–87–94 (*$2.50*)

Change, Process and Events Carol E. Cleland CSLI–87–95 (*Forthcoming*)

One, None, a Hundred Thousand Specification Languages Joseph A. Goguen CSLI–87–96 (*$2.00*)

Constituent Coordination in HPSG Derek Proudian and David Goddeau CSLI–87–97 (*$1.50*)

Lecture Notes

The titles in this series are distributed by the University of Chicago Press and may be purchased in academic or university bookstores or ordered directly from the distributor at 5801 Ellis Avenue, Chicago, Illinois 60637.

A Manual of Intensional Logic Johan van Benthem. Lecture Notes No. 1

Emotions and Focus Helen Fay Nissenbaum. Lecture Notes No. 2

Lectures on Contemporary Syntactic Theories Peter Sells. Lecture Notes No. 3

An Introduction to Unification-Based Approaches to Grammar Stuart M. Shieber. Lecture Notes No. 4

The Semantics of Destructive Lisp Ian A. Mason. Lecture Notes No. 5

An Essay on Facts Ken Olson. Lecture Notes No. 6

Logics of Time and Computation Robert Goldblatt. Lecture Notes No. 7

Word Order and Constituent Structure in German Hans Uszkoreit. Lecture Notes No. 8

Color and Color Perception: A Study in Anthropocentric Realism David Russel Hilbert. Lecture Notes No. 9

Prolog and Natural-Language Analysis Fernando C. N. Pereira and Stuart M. Shieber. Lecture Notes No. 10

Working Papers in Grammatical Theory and Discourse Structure: Interactions of Morphology, Syntax, and Discourse M. Iida, S. Wechsler, and D. Zec (Eds.) with an Introduction by Joan Bresnan. Lecture Notes No. 11

Natural Language Processing in the 1980s: A Bibliography Gerald Gazdar, Alex Franz, Karen Osborne, and Roger Evans. Lecture Notes No. 12

Information-Based Syntax and Semantics Carl Pollard and Ivan Sag. Lecture Notes No. 13

Non-Well-Founded Sets Peter Aczel. Lecture Notes No. 14

A Logic of Attribute-Value Structures and the Theory of Grammar Mark Johnson. Lecture Notes No. 15